planets, potions and parchments

NON-RETURNABLE

planets, potions and parchments

Scientifica Hebraica from the Dead Sea Scrolls to the Eighteenth Century

B. Barry Levy

Published for the Jewish Public Library
by McGill-Queen's University Press
Montreal & Kingston • London • Buffalo

©Jewish Public Library 1990

ISBN 0-7735-0793-0 (cloth)
ISBN 0-7735-0791-4 (paper)

Legal deposit second quarter 1990
Bibliothèque nationale du Québec

Printed in Canada on acid-free paper

Canadian Cataloguing in Publication Data

Levy, B. Barry
Planets, potions, and parchments: scientific
Hebraica from the Dead Sea scrolls to the eighteenth
century

Catalogue of an exhibition, presented by the
Jewish Public Library, held at the
David M. Stewart Museum in Montréal,
from May to September, 1990.
Includes bibliographical references.
ISBN 0-7735-0793-0 (bound). –
ISBN 0-7735-0791-4 (pbk.)
1. Science – History – Sources – Bibliography –
Exhibitions. 2. Medicine – History – Sources –
Bibliography – Exhibitions. 3. Hebrew imprints –
Bibliography – Exhibitions. I. Jewish Public
Library (Montréal, Quebec) II. David M. Stewart
Museum III. Title.
Q105.C3M65 1990   016.509   C90-090230-2

Graphic design and Hebrew calligraphy by:
"LEOGRAPHIC" Effy Givon Designer

Contents

Foreword vii
Preface viii
Acknowledgments ix
Introduction: Of Whirlwinds and Crucibles 1
Chapter One   God and Nature in Ancient Times 9
Chapter Two   Astronomy 19
Chapter Three   Mathematics and Geometry 39
Chapter Four   General Science 51
Chapter Five   Medicine 65
Chapter Six   Science and Religious Ritual 95
Chapter Seven   Geography and Cartography 109
Postscript 133
Glossary 135
Bibliography 136
Index 138

שאלת חלום

כי יש אדם שמגיע ונוגע דבר זה שאני משאלכם עלו
גדול משכב כן וכן אם בשביל וכיך או בשביל
אגדה או כלומר שהוא צריך אמ׳ רב ל׳ לעשות ובר
זה או בהלך בידך זו לעשות סחורה או על דבר שהוא
צריך הראנו מעלמדי חכמ׳ ובא בנסיון ובמעמדי
ישראל שהם מתפללים ואומרים קדוש ואם אין ל׳ כוח
לעשות דבר זה הראנו ברע עבודה זרה וגוים שמתפללם
ובתי מרחצאות שקמעות מעוונפיו בשם והן יו׳ קיפו
וכל הנעש כת׳ כל העט ויהיה תחת מראשותיך ושכב
בבגדים טהורים וקודם שמשכב רחוץ בעים ושכב
ואמת עלדבר שאתה נריך ל׳:

לאהבה

כופו פי פוטי אנו אפפס אנו קמסיר ק׳ כת׳ על שבר או
חמרי ואוכלה׃

אשבעתיך עלך שישדי המעלאך תמנלא אשא וסובב יקום בגופה
דאימעה ימעל ורות צלחי ושי בראשיה ורות פרקין באידן
ובריל אימני ועל׳ ולא יהורמהנך ברגלותי ויהוי מטוק בערפה
יתפין שאל׳ בנה והא ידע מן נפק ומן על ויהוי דוי ואטמעעניו
וי יומ כל יומ׳ בשם אנתיול השם הגדול האל אורא חר
...הקהק ובטש שגר תעשה בור צי׳
... שאות ... שעות כת׳ ולא תמירתא ק׳ כת׳ עלצד ׳׳׳
בתיומט ... ב... אפקוסרא׃

A fragment of a (medicinal?) recipe book including instructions for dealing with dreams and encouraging love, found in the Cairo Geniza. Annenberg Research Institute, Philadelphia, Geniza fragment no. 457 (Cat. no. 128)

Foreword

IRWIN LITVACK, *President, Jewish Public Library*

This beautiful publication will have a unique appeal for every individual who looks at it. Some readers will appreciate it as a memento of a pleasurable aesthetic experience. For others it will satisfy their intellectual curiosity about a previously unexplored area of knowledge. For Jews, it may be a catalyst to achieving a new perspective on their own religion's history. For non-Jews, it may offer a more focused vision of a strand in the densely woven fabric of western civilization.

For myself and others at the Jewish Public Library who have been involved in *Planets, Potions and Parchments* from its conception, this book is an emblem of what the Jewish Public Library has achieved in its seventy-five years of existence. Its publication is the culmination of a year of celebration of seventy-five years of public service in the dissemination of knowledge. From its beginning, as a "home away from home" for immigrant Jews, operating almost exclusively in Yiddish, the Jewish Public Library has evolved into a multi-lingual, multi-service institution, welcoming Montrealers of every age and cultural background to a broadly based hive of learning, leisure activity, and formal programming. The Jewish Public Library is committed to continuing to serve all parts of our pluralistic community with balanced programs, adding new dimensions and meeting additional needs while remaining faithful to our past and to our heritage. While sustaining and nourishing its international reputation as the finest public Jewish research facility in the world outside Israel, the library has pledged itself, especially in recent years, to meet the ever-proliferating demands of contemporary patrons.

In mounting this exhibition, we believe that we have found the perfect vehicle for combining great scholarship with popular appeal. More than that, this particular project, which demonstrates the organic unity of scientific and religious knowledge in Jewish civilization up to the modern era, reflects the very notion that informs our own image of the Jewish Public Library: a place where all forms of knowledge can co-exist in harmony and mutual benefit, contributing to the health and development of an evolving civilization. *Planets, Potions and Parchments* is a fortuitous symbol reinforcing what the Jewish Public Library is achieving.

Knowledge is a fundamental component of the Jewish Public Library, but an equally important component in this endeavour has been our happy collaboration with the people of the David M. Stewart Museum who worked with us to make this dream come true. Indeed, for those of us "behind the scenes," one might have subtitled the exhibition "Plans, People, and Partnerships." Just as this is our first opportunity to offer our treasures to the public at large, so it is the first occasion on which the Jewish Public Library has worked extensively with another institution on a major public project. We have been lucky in our choice, as the Acknowledgments demonstrate.

The Officers and Board of Directors of the Jewish Public Library join me in expressing our sincere good wishes for the success of the exhibition. It is our hope and expectation that *Planets, Potions and Parchments* will have the widest possible appeal, and that visitors of all ages, and from all cultures, will find pleasure and enrichment in its contents.

The final word, fittingly, belongs to one whose life was dedicated to the discovery of scientific knowledge. Roger Bacon, the acclaimed English scientist and essayist wrote:

> From the beginning the Hebrew have been very skillful in the knowledge of astronomy; and all nations have obtained this science as well as other sciences from them.

Preface

MRS DAVID M. STEWART, President, The David M. Stewart Museum

Science: "the possession of knowledge as distinguished from ignorance or misunderstanding."
(Webster's Dictionary, 1973)

The seventy-fifth anniversary of the Jewish Public Library in Montreal is in itself a cause for celebration of a lifetime of promoting the educational and cultural life of our great city and that of the Jewish community in particular. It is an opportunity to reflect on the significant contributions made by Jews throughout their long history here and throughout the world.

The David M. Stewart Museum is proud to participate in this celebration with the presentation of *Planets, Potions and Parchments*, an exhibition that focuses on the role of Jews in scientific thought and development to the end of the eighteenth century, a theme never before undertaken by any museum or institution of learning. It gives the Museum a unique opportunity to exhibit relevant artifacts from its scientific collection, and to bring to Montreal some of the most important books, maps, and illustrations on the subject from major lending institutions in Canada, the United States, Great Britain, Europe, and Israel.

In this catalogue, Barry Levy presents a new interpretation of an area which has long been neglected. For many centuries, Jewish religious studies and science were closely interrelated; it was only in the nineteenth century that these disciplines became separate. New developments in science gave little recognition to early theories and practices that were no longer relevant. At the same time, modern Jewish religious writers tend to ignore the part that the now outdated scientific teachings had played in their predecessors' interpretations of the Bible and the Talmud. This exhibition opens a window on the past and in doing so suggests that ethical concerns over modern scientific discoveries and practice could benefit from the contribution of religion.

The exhibition has provided a rare opportunity for two institutions from different cultural sectors – a library and a museum – to combine talents and resources. It has necessitated a working relationship between the two institutions that we have both enjoyed and valued.

For our modern minds, which understand science as nuclear medicine, space ships to the moon and beyond, and the Concord supersonnic jet, we hope that this catalogue will provide its readers with a singular chance to discover and muse over what science meant to our ancestors – a science filled with astrolabes, whirlwinds, potions, magic, and mystery.

Acknowledgments

ZIPPORAH DUNSKY SHNAY, *Executive Director, Jewish Public Library*

Exhibitions of the magnitude of *Planets, Potions and Parchments* generally require from three to five years preparation. That this exhibition was conceived, prepared, and mounted in something less than two years is a testimony and tribute to the efforts of a dedicated group of people.

The project initially proposed by Hy Goldman was to exhibit selected Hebraica manuscripts and rare books, many fine examples of which are held locally in private collections. His idea was expanded and grew into a collegial effort, finally becoming the focus of the Jewish Public Library's seventy-fifth Anniversary celebrations.

It was our intention that this exhibition, with its never-before-mounted theme of scientific Hebraica from early times to the pre-modern era, both provide the opportunity for scholarship and research and be accessible to the informed layman and the interested public. Parallel to the major theme of the Jewish contribution to the sciences are several concurrent themes: the interrelationship between science and religion, the sharing of knowledge between peoples and the development of printing and book-making. The opportunity to expose children to this rich heritage became one of our primary goals.

The Officers and Board of Directors of the Jewish Public Library recognized the importance and value of this exhibition. Their continuing support, confidence, and encouragement sustained our efforts.

Brad Hill, who was then curator of the Jacob M. Lowy Collection, National Library of Canada, provided the inspiration that led to the choice of the exhibition theme; we thank him for his important contribution and for bibliographic assistance. We also thank Liana Van der Bellen, of the National Library of Canada, for her assistance.

Our exhibition required an accredited museum facility with an experienced organization. It was our good fortune that Mrs David M. Stewart, President of the David M. Stewart Museum, shared our enthusiasm; thus was established a collaborative and productive venture.

From the beginning of this project Mrs David M. Stewart, President of the museum, and Mr Bruce Bolton, its Director, have generously guided me and my staff through the maze of mounting an exhibition. We worked closely with Guy Vadeboncoeur, the curator; Elizabeth Hale and Eilean Meillon, our library counterparts; Guy Ducharme in public relations; and Guy Duchesneau, the museum manager. Support staff, from secretaries to guides, both paid and volunteer, worked unstintingly on every facet of the exhibition, enthusiastically ensuring its success. We owe a special word of thanks to Bruno Donzet of Paris for creating the unique and exciting exhibit design, and to his assistant Monique Rogé for co-ordinating the European loans. With such fine professional consultants as Terry Rempel-Mroz (paper conservator), Siegfried Rempel (environment conservator), and Jill Corner (audio-guide), we were assured an exhibition of international stature.

We express our sincerest appreciation to Nachum Gelber, Q.C., for his invaluable assistance in securing permission to borrow a fragment of the Dead Sea Scrolls from the Israel Museum in Jerusalem. This national treasure, one of the most valuable items in the exhibition, travels only rarely; we are privileged to be able to offer it for public viewing. We also thank Meir Meyer, Vice Chairman of the Israel Museum, and Ruth Peled, Chief Curator, Ministry of Education and Culture, State of Israel.

Few cities can boast of the human resources and breadth and depth of scholarship that it is Montreal's good fortune to have. Of the many internationally acclaimed scholars who do credit to our community, we are fortunate that B. Barry Levy of McGill University, a renowned Bible scholar, accepted the position of guest curator. His efforts to secure relevant and interesting materials took him to several continents, and the scope and magnitude of the exhibition expanded severalfold from the original concept. His book will serve as a lasting record of the exhibition and as a scholarly treatise on the subject. Our thanks are expressed to Ira Robinson, Professor of Judaic Studies, Concordia University, and to Yaakov Rabkin, Professor of History of Science, Department of History, Université de Montréal, for their splendid work in organizing an international symposium on "The Interaction of Scientific and Judaic Cultures".

Gianfranco Silvestro, Director, Instituto Italianio di Cultura, Montreal, was helpful in communicating with officials in Italy. Not the least of our challenges was the number of languages encountered in our contacts with people in various countries.

We trust that this beautiful catalogue will serve not only as a permanent historical record of the

A fragment containing "a great secret for intercourse, particularly for a first sexual encounter, especially for one who has difficulty achieving an erection." The Wellcome Institutue Library, London, MS Heb. A 25 (Cat. no. 127)

library's seventy-fifth Anniversary and the exhibition but also as a reference tool for research and scholarship. We express our appreciation to Emmanuel Kalles, a former President of the Jewish Public Library, for his invaluable assistance in arranging this publication with McGill-Queen's University Press. And to the Press, a sincere word of appreciation for their professional care and concern.

We acknowledge all the people without whose help *Planets, Potions and Parchments* would have been impossible to assemble; in particular we thank: Nigel Allen of the Wellcome Institute for the History of Medicine; Esra Kahn, Librarian of Jews College; Adrian Roberts and R.C. Judd of the Bodleian Library; A.A.E.E. Ettinghausen, OBE of the Montefiore Endowment; Emmanuel LeRoy Ladurie, administrateur général de la Bibliothèque nationale, Paris, and his staff; Father Leonard Boyle, of the Biblioteca Apostolica, and Cardinal Agostino Casaroli, The Vatican; Franca Arduini of the Biblioteca Universitaria, Bologna; Marcello Pavarani of the Biblioteca Palatina, Parma; Pearl Berger, Dean of Libraries, Yeshiva University; Menahem Schmelzer and Rabbi Jerry Schwarzbard of the Jewish Theological Seminary, New York; Kenneth A. Lofh, Head, Rare Books Room, Butler Library, Columbia University, New York; Leonard Gold, Dorot Chief Librarian of the Jewish Division, New York Public Library, New York; Vivian Mann, Curator of Judaica, The Jewish Museum, New York; Mary Wyly, Associate Librarian, the Newberry Library, Chicago; John Chalmers, Head Librarian, the Harry Ransom Humanities Research Centre, University of Texas, David M. Goldenberg, Aviva E. Astrinsky, Michael Terry, and Mordecai Ostwald of the Annenberg Research Institute, Philadelphia; Lisa Kightlinger and Eileen Devinney, of the University Museum, Philadephia, Pennsylvania; Arnona Rudavsky and David Gilner of Hebrew Union College, Cincinnati; E. Keall, curator, West Asian Department, Royal Ontario Museum; Richard Virr of the Rare Books Room, McLennan Library, Faith Wallis, Osler Library, and Barbara Lawson, Redpath Museum, McGill University; Reverend William Klempa, Principal, and Dan Shute, of Presbyterian College, McGill University; Terry Lightman, Museum Curator, The Shaar Hashomayim Congregation, Montreal; Y. Tzvi Langerman, for his helpful corrections and suggestions; Joel Linsider, for editorial assistance; Stefani Novick, research assistant; Rose Lenkov, for assistance and support far beyond the call of duty.

I express our profound appreciation to all the lending institutions, as well as to the individuals who are sharing items from their collections. Permission to reproduce photographs of the books and manuscripts is gratefully acknowledged.

We gratefully acknowledge assistance, financial and other, from Communications Canada, Museum Assistance Program/Programme d'appui aux musées, and Travelling Exhibition Insurance Program/Programme d'assurance des expositions itinérantes; Ministère des Affaires culturelles; Ministère de l'Énergie et des Ressources; Ministère des Communautés culturelles et de l'Immigration, Gouvernement du Québec/Government of Quebec; Le Conseil des Arts de la Communauté urbaine de Montréal; Commission d'initiative et de développement culturels (CIDEC), Ville de Montréal; and the Jewish Community Foundation of Greater Montreal.

Bryna Garmaise and her committee worked tirelessly to establish a Speaker's Bureau to amplify the exhibition; guides were trained and public lectures planned. Mona Polachek, Barbara Kay, and Victor Abikhzer also brought their expertise to bear in these areas. Barbara Kay's contribution to the editing process was, as always, energetic, witty, and intelligent.

Eva Raby, Head, Norman Berman Children's Library of the Jewish Public Library, provided input to the overall planning and organization; specifically she brought her outstanding talents to the task of ensuring the participation of the schools through class visits, and creating the pedagogic kits and teachers' seminars.

Claire Stern, Head of Public Services, coordinated the Speakers' Bureau and the Public Programs; she undertook, in her usual prompt and meticulous manner, the tasks of establishing standards for translations and transliterations used in the text. Ron Finegold provided able assistance.

Glen Rotchin gave cheerful help in countless ways, and for many hours; Donna Shoham assisted with public relations, as well as ably managing ticket sales. The public's response attests to their capabilities.

Belline Litman cheerfully spent many hours reviewing and editing French text; her expert advice was invaluable.

I also thank Charles Kless, my Administrative Assistant, who relieved me of many duties and allowed me to concentrate on the exhibition.

To the entire staff of the Jewish Public Library for cheerfully accepting additional tasks, for maintaining their usual high standards, and for their continued support, I express my heartfelt thanks.

Our appreciation must be expressed to Allied Jewish Community Services (AJCS) for their encouragement and moral support; in particular we thank the President, Maxine Sigman and the Executive Director, John Fishel. Raphael Assor, Director of Governement Relations, provided excellent liaison and advice.

To Robert, who is still my husband in spite of my near total absence on the domestic front within the past year, I want to publicly acknowledge his support encourasgenent, and unfailing good humour throughout this project.

To all those whom I may have inadvertently omitted, I beg their indulgence and forgiveness.

On a personal level, the satisfactions I enjoyed while working on this exhibition more than compensated for the sabbatical manqué. The challenges inherent in bringing together the many disparate elements of this complex project, the intellectual stimulation, and the pleasures of working with diverse people were a source of great pleasure, pride, and personal growth.

Lenders to the Exhibition

We are grateful to the following individuals and organizations and their governing bodies who loaned items to *Planets, Potions, and Parchments* and gave permission to describe and reproduce the items in this catalogue.

The Wellcome Institute for the History of Medicine, London, England
Jews College, London, England
The Bodleian Library, Oxford, England
The Montefiore Endowment at Jews College, London
La Bibliothèque nationale, Paris, France
Biblioteca Apostolica, The Vatican
Biblioteca Palatina, Parma, Italy
Biblioteca Universitaria, Bologna, Italy
The Israel Museum, Jerusalem, Israel
The Newberry Library, Chicago, Illinois
Yeshiva University, New York, New York
The Jewish Theological Seminary, New York
Columbia University, New York, New York
Annenberg Research Institute, Philadelphia, Pennsylvania
The University Museum, Philadelphia, Pennsylvania
Hebrew Union College–Jewish Institute of Religion, Cincinnati, Ohio
The National Library of Canada, Ottawa
The Royal Ontario Museum, Toronto
McGill University, Montreal
Presbyterian College, Montreal
The Jewish Public Library, Montreal
Ofra Aslan, Montreal
Yehuda Elberg, Montreal
B. Barry Levy, Montreal
Dr and Mrs Julius Pfeiffer, Montreal

Advisory Committee

Mrs David M. Stewart, President of the David M. Stewart Museum
Irwin Litvack, President of the Jewish Public Library

Bruce Bolton
Bryna Garmaise
Aron Gronshor
B. Barry Levy
Eva Raby
Irving Rackover
Gina Roitman
Zipporah Shnay

Honorary Committee Gala Preview

Co-Chair:
Marjorie Bronfman
Liliane Stewart

Michel Belanger
Leonard Cohen
Norma Cummings
Gloria Davis
Victor Goldbloom, Q.C., LL.D.
Lily Ivanier
Johanne Jacobs
Melanie King
Alan Klinkhoff
Marcel Knesht
Bernard Lamarre
Murray Lapin, Q.C.
Harvey Levenson
Nicki Papachristidis
Dorothy Reitman
Jacqueline Simard
H. Heward Stikeman, Q.C., LL.D.

Gala Preview Planning Committee

Anna Gonshor, Chair

Norma Cummings
Shirley Garfinkle
Pearl Lighter
Alison Lyons
Mona Polachek
Zipporah Shnay
Doris Weiser

Public Programming

Bryna Garmaise, Chair

Victor Abikhzer
Pierre Anctil
Batia Bettman
Mervin Butovsky
Rachel Cohen
Miri Flakowicz
Regine Frankel
Anna Gonshor
Edna Janco
Manny Kalles
Barbara Kay
B. Barry Levy
Patti Litvack
Marilyn Nayer
Alexis Nouss
Barbara Pascal
Mona Polachek
Irving Rackover
Ira Robinson
Robert Wiseman

Speakers Bureau – English Language

Barbara Kay, Chair

Bryna Garmaise
Glen Rotchin
Donna Shoham
Claire Stern
Robert Wiseman

Speakers Bureau – French Language

Victor Abikhzer, Chair

Pierre Anctil
Daniel Bonneterre
Shalom Cohen
Suzanne Dadoun
Louise Druckman
Philippe Elharar
Regine Frankel
Elias Levy
Yossi Levy
Daniel Malka
Jimmy Muyal
Alexis Nouss
Moshe Shalom

Educational Materials

Eva Raby, Chair

Batia Bettman
Jerry Dunn
Miri Flakowicz
B. Barry Levy
Honey Fox Moskowitz
Barbara Pascal
Mona Polachek
Judy Stein

Guides Initiation

Mona Polachek, Chair

Louise Druckman
Regine Frankel

פרק ראשון

בו יבואר מספר העגולים תמונתם ומקומם כדעת התוכנים הקדמונים ובפירושם:

בכדור השמים התוכנים הקדמונים כונים מספר י' עגולים ואלה הם: עגול משוה היום: עגול המזלות: עגול עמידת השמש: עגול קו השווים: עגול רצי היום: עגול האופק: עגול גדי: עגול סרטן: עגול צפוני: עגול דרומי: ששה עגולים הראשונים נקראים גדולים: ארבעה האחרונים נקראים קטנים: בערך הגדולים: עגול גדול כל עגול אשר יורשם בכדור וירחתך הכדור אשר יעבור על מרכזו לשני הצאים שוים: והעגול הקטן הוא העגול אשר יהיה קטן מזה והוא יחתך הכדור בחלקים בלתי שוים כי החלק האחד יהיה יותר קטן מהחצי האחר ויהיה נתחייב בהכרה שלא יעבור על המרכז ויטה ממנו וכמו שיראה בכדור המוגשם לפניך

דף מח

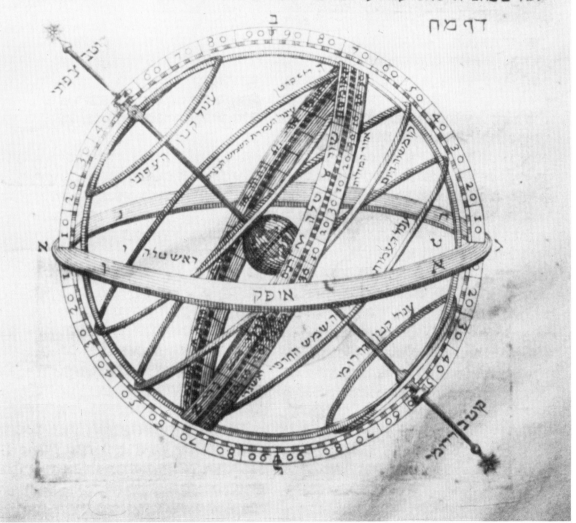

An armillary sphere, Tobias Cohn, *Ma'aseh Tuviah*, Venice, 1708. National Library of Canada, Jacob M. Lowy Collection (Cat. no. 73)

Title page from Benjamin of Tudela's *Masaʿot Shel Rabbi Binyamin*, Ferrara, 1556. National Library of Canada, Jacob M. Lowy Collection (Cat. no. 169)

Introduction
Of Whirlwinds and Crucibles

God and Nature

Job, the pious hero of the biblical book that bears his name, was challenged by Satan (with God's approval) in a test of his faith. After seemingly endless tragedy, including loss of his wealth, the deaths of his children, and continuous physical suffering, Job demanded that God reveal to him his sins, the justifications of his presumably divinely ordained torture. Unmoved by appeals to justice or by the psychological needs of His devoted servant, God declined to answer Job's request. Instead, He confronted him from a whirlwind (a tornado or a storm) with a lengthy series of taunts, including:

> Where were you when I built the earth's foundations?
> Did you ever command the day to begin?
> Have you gone to the depths of the sea?
> Do you know the laws of heaven?
> Can you raise your voice to the clouds?
> Can you make lightning go?
> Do you know when mountain goats give birth?
> Do you give the horse strength?
> Does the eagle fly at your voice? ...
>
> Job 38–39

Job's meek response,

> ... Therefore, I spoke but did not understand,
> Things too wondrous for me I did not know ... (42:3),

conceded the strength of God's position. God's mastery of these and other natural phenomena set Him apart from man. Having created both man and nature, God epitomized the knowledge and power that placed Him above and beyond the levels of ignorance and impotence exemplified by Job and, by extension, the rest of the human race. Taken to its radical conclusion, this assumption leaves God the Creator and Master, and people insignificant creatures whose only real knowledge is of their insignificance and whose only grace lies in suffering under the divine master's harangues.

Though this position did influence some religious thinkers, it was not adopted thoughout the Bible and did not become a normative Jewish idea. Nor, on the whole, did Western civilization identify with it. In a symbolic sense, people reacted to it by trying to learn what only God seemed to know; the evolution of this quest for knowledge of nature – the history of science – has been the human response to God's intimidation of Job.

To be sure, scientific awareness predates Job. The ancient Sumerians and Egyptians were keen observers of nature. Solomon, a biblical sage of legendary proportions, is described as a master of natural science – having knowledge of all types of trees, animals, birds, insects, and fish (*I Kings* 5:13-14). The ancients possessed much medical and astronomical knowledge; skulls discovered at biblical and pre-biblical sites, for instance, conclusively demonstrate their ability to perform successful cranial surgery. The devising of medical treatments and quarantine, the creation of calendars, predictions of astronomical phenomena, the development of mathematical sophistication, and the awareness of procedures that could enhance agricultural productivity all underscore ongoing attempts throughout Antiquity to master nature and the environment. But these were often accompanied by beliefs to the effect that political and military successes or failures were a diety's conscious choices, that the presence or absence of rainfall was divine response to human behaviour, and that human illness was punishment for violating religious taboos. Concomitant beliefs often assumed divine abilities to do whatever was desired. Divine interference in nature was deemed to be a possible, if not a frequent, event.

The Bible repeatedly portrays God as above nature and in control of it. This position stood in contrast to ancient pantheistic paganism, which perceived divine and semi-divine forces in all objects and natural phenomena, and offered a litany of rites and observances to manipulate them, or a least to attract their positive aspects and deflect their negative ones. The Bible offered the notion that the universe had been created by one God, who could control its forces and use them to serve His will, in some cases as rewards and punishments for His creatures. Occasionally, He even changed the course of what appears as nature to achieve these goals, as when He sent the plagues against the Egyptians through Moses, delayed sunset so that Joshua could complete a battle, captured Jonah in the belly of a fish after he had been trapped in a storm, and defeated the four hundred Baal priests through a heavenly fire brought down at the demand of Elijah.

On Miracles

While the Bible seems quite comfortable with

these stories, post-biblical thinkers were often troubled by them. Philo of Alexandria (first century CE), the best-known Jewish thinker of Hellenistic Egypt, asserted God's ability to influence the world by directing the natural forces that He had created. But shortly thereafter, the rabbis in Israel taught that many objects and events seemingly not in conformity with nature had been created during the final moments of the sixth day of creation. They thus suggested that these apparent exceptions to nature were, in themselves, part of it, not deviations or alterations after the fact. A few centuries later, Genesis Rabbah went even further, articulating a principle that accommodated virtually every possible change. It suggested that God contracted with each thing He created that it would be available to perform the unusual acts He might request in the future. Thus, all matter was enlisted in the service of God and prepared to alter its natural course.

Medieval philosophers wrote extensively about miracles, as the unusual events described in these passages came to be known. Some perceived them as evidence of divine concern for mankind and direct interference in the course of human existence; miracles thus proved God's existence and His involvement with people. Such thinkers were prone to read biblical miracle stories as accurate historical reports, though they were not necessarily of one mind on how to understand exactly what had occurred.

Others agreed that the Bible would not relate untruths, but suggested that the truths of the miracle narratives might fall outside the historical realm. In other words, the point of an account of a miracle could be to teach divine protection of Israel, to warn of the impending punishment of evildoers, or to glorify the Creator. Some representatives of this group went so far as to claim that biblical narratives were written as miracle stories to captivate and educate the naive and ignorant, but that sophisticated readers would appreciate the truth behind them and not be misled by their fabulous aspects. In this case, an allegorical reading of each story – which offered an abstract concept, not historical information, as its main point – would be required.

Other approaches to explaining divine interference in nature were suggested as well, but one thing remained clear: either God could be credited with the ability to change natural law or He could not. In the former case, virtually anything became possible; in the latter, the particular biblical passage had to be either interpreted in light of what was accepted as possible according to non-biblical thinking or else be deprived of any serious authority.

Problems inherent in these positions, coupled with scientific skepticism about religious teachings or the pious assumption that God and His actions are beyond human understanding, have fostered the anti-scientific attitudes popularized by some religious writers. Some interpreters even used them to justify augmenting miracles described in the Bible almost at will, substituting, through midrashic fancy, miracle on top of miracle for some scientifically defensible biblical reports.

Judaism and Science

Pre-modern writers, for the most part, believed in God and the Bible. Therefore, though the potential for conflicts with emerging scientific knowledge was great, the commitment to resolving the conflicts was strong. The results of this commitment are evident in the well-documented presence of scientific discussions in religious books and in the fact that many important Hebrew scientific works were composed by, with the support of, or at the request of the same rabbis who served as the nation's religious leaders.

Frequently, the authors of the most important medieval Jewish tracts on mathematics, astronomy, geography, and medicine are well known as rabbinic writers. Maimonides, Moses Isserles, Elijah Mizrahi, Abraham Ibn Ezra, Gersonides, and Mordecai Jaffe are famous as Bible commentators or halakhic authorities; their names are well known to all who dabble, however casually, in classical rabbinic literature. Their discussions of scientific matters are less well known, but, in some cases, they were of universal significance, not merely Jewish interest. It may be somewhat imprecise to call them scientists and their works scientific, because application of the term to medieval thinkers is anachronistic, but this exhibition highlights their contributions to what we now call the sciences, and no more satisfactory term is available.

Forceful pressures to convert were exerted on Iberian Jews from the end of the fourteenth century until the expulsion a century later. So many of these people and their children, known variously as Marranos, conversos, new Christians, and secret Jews, advanced to positions of power in the church and government that major efforts were undertaken to certify and preserve bearers of pure Spanish blood. Indeed, it often seemed impossible for pure-blooded Iberians to resist the collective assault on knowledge, wealth, and power waged by this important group. Both Jewish and non-Jewish leaders vigorously debated the status of the Marranos, many of whom returned to Judaism when given the opportunity to escape to Amsterdam, Venice, or other friendly cities. But their access to the universities and to the most advanced scientific knowledge of the day ensured their impact on cartography, astronomy, medicine, and other sciences; in the context of recent Jewish scholarly trends that acknowledge their Jewishness, they, too, deserve a place in the exhibition.

Jewish and Marrano contributions were shared fully with the non-Jewish world. Jews often wrote in Arabic, Latin, or the other accepted languages of scientific expression, and translated works into Hebrew, thereby ensuring that the essential compositions by learned Christians and Moslems would be available to their co-religionists who were less fluent in these languages. Hundreds of important scientific works were translated from Arabic into Hebrew, and often from Hebrew into Latin; similar exchanges were routine for works composed in Hebrew. Indeed, one may readily document both the international and interconfessional nature of the medieval scientific endeavour and the important role in its development played by Jewish writers and translators.

The multi-faceted education of medieval times encouraged mastery of all knowledge as it existed, and therefore doctors of sacred law (Moslem, Christian, or Jewish) could be expected to study medicine, astronomy, and the other scientific disciplines. In fact, mastery of halakhah (Jewish religious law), the professional requirement of every rabbi, necessitated extensive knowledge and understanding of astronomy, the internal and external anatomy of all kosher animals, the theories of combustion and causality, all sorts of botanical information, and many aspects of gynaecology, pediatrics, and internal medicine. Geometry was needed for calculating the sizes of irregular plots of land, and familiarity with the technicalities of mathematics was so necessary for rendering decisions in inheritance claims that compendia of these laws sometimes began with math lessons. Astronomical works ostensibly devoted to explaining the workings of the calendar were often introduced by a lengthy essay on geometry, and the laws of probability were an important element in the codification of many problems associated with the dietary regulations.

The need to know specific aspects of these sciences assumed the integration of scientific and religious concerns; the intellectual abilities and accomplishments of the people who were involved virtually ensured their additional curiosity, thorough study of the fields, and, in some cases, original contributions to them. But even if these factors were deemed of secondary importance, the vast number of astronomical, biological, mathematical, and medical discussions in the Talmud required that any serious talmudist be conversant with much scientific information, and careful examination of the medieval commentaries, an essential aspect of all Talmud study, shows that the scientific data had advanced beyond any limitations inherent in its local Babylonian origins.

Nothing illustrates the relationship between the sciences and Judaism more clearly than the sheer numbers of Hebrew manuscripts and books on the sciences. Thousands of Jews contributed to the study of the sciences, and many wrote scientific books. Many libraries possess hundreds of Hebrew manuscripts, each of which is a collection of scientific compositions.

Many of these have never been published, and very few of those that have been exist in scientific editions. This situation highlights a paradoxical trend that has dominated Jewish studies for the past several centuries. As late as the eighteenth century, writers were composing synthetic works in which religion and science informed and influenced each other. As the sciences drew away from religion, and religion from the sciences, this pattern ceased to exist, and several alternatives emerged.

The religious community tended to ignore scientific concerns and to focus on religious and spiritual issues. It is inaccurate to attribute this trend only to the emergence of Hasidic Judaism, because some Hasidic writers, such as Rabbi Gershon Henoch (who, after much scientific research, reinstituted the use of blue threads in the *tzitzit*) and Rabbi Solomon of Helm (who had a strong interest in mathematics), did contribute to the Jewish scientific literature. But the mystical atmosphere fostered by many Hasidic groups left little room for independent scientific thinking.

As Hasidism was developing into a leading religious movement, the scholarly study of Judaism, which explored many aspects of the religion's evolution in an attempt to understand its history literature, beliefs, and practices, emerged as an important intellectual priority. But while it attempted to be scientific, it was not particularly interested in science. *Wissenschaft des Judentums,* "the science of Judaism," as this intellectual endeavour was known, was, itself, not overly concerned with *Wissenschaft;* it had its own agenda, dominated by a deep concern for the intellectual study of Judaism and often by a desire for religious reform.

The legitimate institutional heirs to the medieval commitment to integrating religion and science are, ironically, Orthodox — Yeshiva University and several Israeli schools of higher learning that describe themselves as both scientific and religious. Yeshiva University's motto, *Torah uMadaᶜ* ("Torah and Science") has emerged from decades of ideological scrutiny as a call for the synthesis of religious and scientific studies. How it fulfills this goal and how that fulfillment differs from the medieval model on which it is based are beyond our present concern, but similarly minded institutions that teach commitment to these potentially conflicting systems of thought are rare indeed.

Science for Its Own Sake

Much of the scientific study undertaken by Jews in ancient and medieval times was motivated by practical concerns — religious, medical, or agricultural. But the exploration of science for its own sake also was far from unusual, as can be seen from the range of theoretical discussions that found their way into the literature. Inevitably, the number of non-Jewish scientific thinkers far surpassed the number of Jewish ones, and their collective contribution was therefore far greater. This led to some later perceptions of science as foreign to and even as conflicting with Jewish interests, which sometimes tended to draw Jews away from science and to leave the impression that they did not contribute measurably to its development.

In some cases, the patent for a particular tool or notion properly belongs to a Jewish writer, but it remained for modern historians to ensure proper attribution. The function of the Fallopian tubes, called *shevil* ("path") in Hebrew and named after Fallopias, who discovered, in the sixteenth century, that they served as paths for the ova, is described by Maimonides in *Hilkhot Issurei Bi'ah* 5:4, though it is possible that he depended on earlier medical sources. That hemophilia was carried through the female's family was discussed in the Babylonian Talmud. And credit for the invention of the Jacob's Staff (properly Gersonides') was often attributed to others. In short, science was a respected international activity in which Jews shared, from which they learned, to which they contributed, and which, in many cases, had a profound impact on both their physical and spiritual lives.

Papers and Parchments

The books and manuscripts assembled here can be examined from a number of perspectives. Each is part of the history of science and contains interesting data on the pre-modern evolution of the quest to understand "the world and fullness thereof." Its contribution to the modern understanding of the world may be minimal or nil, but this is not necessarily indicative of its historical impact, though some items – for instance, most of the folk cures and amulets – were never of great importance.

Because many of the books were composed by rabbis or by scholars who wrote with their religious leaders' support or encouragement, they also should be explored as components of Jewish religious history. Those who wish to derive contemporary religious guidance from earlier Jewish writers may learn much about the integration of science and religion from some of their attempts at synthesis, even if the scientific contents are outdated. Those opposed to such endeavours will find support for ignoring such works and their intellectual model in the fact that many compromises and reinterpretations suggested by earlier scientists have been discredited by later ones. Of course, many of the compositions have had relatively little impact on the development of modern science, while others have had equally little impact on the teachings of contemporary Judaism.

Because few of the texts assembled are autographs, virtually all of them exemplify the efforts made to copy and recopy the literature preserved by Jews for use in scientific study. In some cases, they also provide important variant readings in marginal or inter-linear notations, and a few works are presented in both manuscript and printed versions, which allows a comparison of these two forms of transmission.

The exhibited items also offer a good sample of the scholarly languages used by Jews during the last thousand years, including Hebrew, Arabic, Latin (which was the language of European scholarly discourse and not reserved for the church), Spanish, Aramaic, Portuguese, German, and Yiddish. They also portray the history of writing, the evolution of both the Semitic and Western scripts, the types of writing materials in use — ceramic bowls, papers, and parchments — and the resultant types of amulets, manuscripts, and printed books.

The ancient and medieval sciences were traditionally subdivided into four or seven categories, and Renaissance writers occasionally added others, such as economics and printing. The modern sciences include a host of previously unfathomed specialties and sub-specialties, to which historians of science have sometimes added history, philology, linguistics, and other areas often classified as part of the humanities and social sciences. But, simply put, most of the items discussed below can be classified as part of either the physical or life sciences. The former are dominated by astronomy but include formal cosmologies, mathematics, and geometry, applications of all three to determining the calendar, and meteorology. The latter are largely concerned with medicine and general biology, and also include related rituals such as kosher slaughter and circumcision.

All of the works listed as general science fit into both of these categories, and many of the specialized works assigned to the seven chapters in this catalogue contain discussions of several areas and could have been placed differently. The science-based religious rituals, grouped, for the sake of convenience, in an independent chapter, could also fit neatly into one or another scientific category. Though it has seemed best to group the various books and manuscripts into several chapters and a number of smaller sub-units, many items could easily serve as examples of several fields, and, in several cases, photographs of items displayed in one chapter have also been added to another.

While science is largely descriptive, pre-modern writers often sought to find meaning in various natural phenomena and scientific facts. Thus, astronomy was applied to human circumstances through astrology; mathematics became the basis of a sophisticated system of numerology; and medicine was helped along through kabbalistic amulets that complemented biology. The practice of bloodletting was considered medicine; determining when it was efficacious to perform the procedure was controlled by astrology.

Jewish books about science can be studied from several different perspectives. Here, much attention is given to the internal Jewish significance of the material, and not only to its relevance to the history of science. This reflects my personal interest in the history of Judaism, and recognizes the importance of the classical concern for the integration of religion and science.

It is unquestionably true that many of the leading medieval rabbis were well trained in the sciences. While some challenged the use of philosophical-scientific works that necessitated a radical re-evaluation of traditonal religious thinking, many others did not. During the Renaissance, in particular, the sciences were well integrated with religious concerns, and the literature is replete with examples of kabbalistic scientists, pious writers who accepted and worked to synthesize the teachings of both tradition and science. During the late pre-modern era this attitude changed, and its implications for the history of Judaism have been profound and pervasive.

I would estimate that about thirty percent of the men whose commentaries appear in the Amsterdam 1724 Rabbinic Bible *(Qehillot Moshe)* also contributed to the study of the sciences. However, about ten percent of those whose writings were included in the Romm edition of the Babylonian Talmud did so. This highlights one difference between eighteenth and nineteenth century approaches to the interpretation of sacred literature: the latter favoured writers with fewer scientific credentials. But it also points to a major intellectual split in the religious community. With some marked exceptions, talmudists tended to study and teach Talmud in isolation or through more traditional approaches, and those who did not are less represented in what has become the standard Talmud edition. Biblicists, in contrast, were more involved as a group in the study of philosophy and the sciences, and their approach to the Bible was therefore informed by these disciplines.

This created a schism that continued to grow in subsequent years and produced a situation in which the Talmud has become virtually the sole concern of Orthodox Jewish religious education. The Bible, philosophy, and science (as well as history, literature, grammar, and other traditional subjects) have all been relegated to marginal status. It is not immediately clear whether Bible study was neglected primarily because those most devoted to it were also committed to philosophy and science, but that seems to be a part of the reason. In many circles it offered a competing rather than a complementary approach to the rabbinic world view. Perhaps philosophy, science, and the Bible appeared to be essentially one interrelated, inherently problematic discipline, whose free spirit and critical goals were uncontrollable and best shunned.

Using the Catalogue

The entries describing the exhibited items contain two parts. Following the item's catalogue number is essential bibliographic information. This includes the title, the author, date and place of publication, the publisher, and the name of the lending institution. Manuscripts are provided, in the place of publication information, with available data about the scribe, script, and place and date copied. For the latter, I have relied heavily on the published catalogues of the collections, which are sometimes incomplete, especially when the date and place of production must be determined solely through paleography. These library collections are listed in a special section of the bibliography, and the manuscript number follows the name of the lending institution.

The entries also provide brief biographical and historical information about the author and his work. Since many writers contributed to more than one scientific area, the biographical information is usually found in the first entry, and subsequent references are marked with an asterisk, which indicates that the person is discussed elsewhere in the catalogue. The precise location(s) can be determined through the index.

Some writers discussed in this catalogue are the subjects of lengthy monographs or books, but others are virtually unknown; sometimes, the available biographical sketches repeat essentially the same few facts. While I have verified much, in many cases full verification has been impossible; despite the improvements that additional research might afford (or that experts on the history of science might make), I have not had sufficient time to attempt the task between being asked to serve as Guest Curator and the required date for completing the catalogue, some six months later. I beg the reader's indulgence for sometimes having gleaned rather than harvested my own crop; I pray that a sufficient amount of grain has been collected, nonetheless, and that a minimum of chaff has remained after the winnowing. Those who wish to explore individual writers or topics further may be initially guided by the bibliography, which includes references for general works and for almost every specific catalogue entry.

Psalm Scroll from Masada. The Israel Museum, The Shrine of the Book, Jerusalem (Cat. no. 4)

Chapter One

God and Nature in Ancient Times

הידעת חקות שמים אם
תשים משטרו בארץ:

איוב ל״ח, ל״ג

> The voice of the Lord is over the waters...
> The voice of the Lord is powerful
> The voice of the Lord is glorious
> The voice of the Lord smashes cedars...
> The voice of the Lord cuts like fire
> The voice of the Lord shakes the desert...
>
> Psalm 29:3–8

> And God said [to Elijah], "Come out and stand on the mountain before the Lord."
> And the Lord passed by; and there was an extremely powerful wind that split mountains and smashed boulders. The Lord was not in the wind.
> And after the wind there was an earthquake; and the Lord was not in the earthquake.
> And after the earthquake, there was a fire; and the Lord was not in the fire.
> But after the fire, a soft, low voice ...
>
> I Kings 19:11–12

Job 38:4–15 with the commentary of Gersonides, Naples, 1487. National Library of Canada, Jacob M. Lowy Collection (Cat. no. 1)

The first page of Job from *Arbaʿah VeʿEsrim*, with the commentaries of Nahmanides and Abraham Farissol, Venice, 1516–17. National Library of Canada, Jacob M. Lowy Collection (Cat. no. 2)

Bible. Job. Hebrew

תנ״ך. איוב

[Iyyov]
Naples: Joseph ben Jacob Ashkenazi Gunzenhauser and Samuel ben Samuel of Rome, 26 September 1487
National Library of Canada, Ottawa (Jacob M. Lowy Collection)

Widely known as one of the most engaging books in the Bible, Job is also at the pinnacle of the ancient wisdom literature, exemplified as well by the "Babylonian Theodicy" and the Akkadian work *Ludlul bel nemeqi*. Its hero's attempts to understand his plight have stimulated hundreds of explorations of the righteous sufferer and have captured the minds and souls of well over twenty centuries of readers and scholars. The history of the book's interpretative literature is, to a large extent, the record of the greatest Jewish and Christian thinkers' attempts to confront this baffling aspect of human existence.

Job's demand that God reveal the reasons for his suffering was met with a series of questions about nature that forced him to concede his inability to fathom God's world. This answer, if one may call it that, assumed that knowledge of nature was a prerequisite for knowing God. One could not hope to understand His ways without first mastering those of the world He had created. The study of nature is thus the starting point for both science and theology, for physics and for metaphysics. For our purposes, the divine challenge to Job to master nature before complaining about his personal situation marks the symbolic begininng of the scientific quest.

This copy of Job is part of a very early edition of the Hagiographa, the third and final section of the Bible. Job is the only section in the volume accompanied by the commentary of *Gersonides (1288–1344), a popular Bible interpreter and important medieval scientist. The commentary was generally regarded as the best classical analysis of the book, although some writers took issue with Gersonides' position, which follows *Maimonides' in seeing divine providence as limited in the sublunar world and applicable only to those with highly developed intellects.

Bible. Hebrew

ארבעה ועשרים

[Arbaʿah VeʿEsrim]
Venice: Daniel Bomberg, 1516–17
National Library of Canada, Ottawa (Jacob M. Lowy Collection)

The early rabbinic Bibles were large folio editions of Scripture accompanied by the commentaries of leading rabbinic thinkers. In Venice, 1516–17, the first to appear, each biblical book was accompanied by a single Aramaic translation and a single commentary, usually that of Rabbi Solomon ben Isaac (Rashi) or Rabbi David Kimhi (Radak). The most noteworthy exception to this pattern is Job, which was accompanied by two commentaries, those of Moses ben Nahman (Nahmanides) and Abraham *Farissol.

Nahmanides, whose literary and intellectual talents far surpassed those of most great medieval figures, distinguished himself as a talmudist, Torah commentator, mystic, and halakhist. His approach to Bible interpretation in general, and to Job in particular, differed radically from that of *Maimonides, his predecessor by some seven decades. Maimonides' analysis of Job was predicated on the talmudic opinion that the events in the book were to be understood allegorically, and on his own observation that the characters represented the main schools of medieval philosophical thought. Nahmanides, on the other hand, saw Job and his comrades as historical persons who reflected authentically Jewish – even kabbalistic – thinking.

Abraham Farissol, an important Italian figure, followed the allegorical approach to Job established by Maimonides, while incorporating various aspects of *Gersonides' interpretation as well. But he did not accept his predecessors' radical limitation of divine providence to the intellectual elite. He and Nahmanides offered complementary approaches that avoided the more radical position of Maimonides and Gersonides.

Rabbinic Bibles were published five more times during the sixteenth and seventeenth centuries, mostly in Venice, always following what became the standard layout and including the standard selection of interpretations: in the case of Job, the Aramaic translation (Targum) and the commentaries of Abraham *Ibn Ezra and Gersonides. The commentary attributed to Rashi was first included in the margins of the 1546–48 edition, and was retained in all subsequent ones.

The first page of Job, *Qehillot Moshe*, Amsterdam, 1724–27. Jewish Public Library, Montreal (Cat. no. 3)

Bible. Hagiographa. Hebrew

קהילות משה

[Qehillot Moshe]
Amsterdam: Moses Frankfurter, 1724–27
Jewish Public Library, Montreal

The first Rabbinic Bible to deviate seriously from the model followed in the second through sixth editions began to appear in Amsterdam in 1724. True to the editor's goal, this set, *Qehillot Moshe*, was a collection of important and sometimes otherwise unavailable commentaries, which was intended to appeal to both the Ashkenazic and Sefardic communities that flourished in Amsterdam at that time. Unlike its predecessors, this Rabbinic Bible provided an anthology of nine medieval and renaissance compositions that grappled with Job, including the works of *Nahmanides, Seforno, Ibn Yahya, Yavetz, Duran, and one attributed to Rashi.

It is significant that a number of commentators on Job in this edition also contributed to the study of medieval science, as this shows both that scientists were writing about Job and that accepted Bible commentators were leading scientists. Abraham *Ibn Ezra was an important mathematical and astronomical figure; Abraham *Farissol contributed to the study of geography; and *Gersonides was an international leader in astronomy and mathematics. *Maimonides, known for his astronomical, mathematical, and medical work, did not compose a commentary on Job but did discuss it in his *Guide for the Perplexed (III, 22–23)*, and thereby influenced a number of important commentators.

Bible. Psalms. Hebrew

תהילים

[Tehillim]
Judea: First century
The Israel Museum, The Shrine of the Book, Jerusalem

Occasional discoveries of ancient texts in the Dead Sea area are reported in Jewish, Moslem, and Christian writings of the past two thousand years, but no archaeological find has received as much attention as the discovery of the collection of books known as the Dead Sea Scrolls. During the past five decades, both casual and professional searchers have continued to recover fragments of hundreds of ancient texts, including sectarian works, prayers, legal documents, Bible translations and interpretations, and the most ancient copies of biblical books in existence. Many are invaluable for understanding the evolution of Judaism and its sacred literature.

The biblical texts discovered in the Dead Sea area (whose climate is favourable to preserving ancient parchments) date from pre-Christian times to the second century and fall into several broad textual types. Some, like the scroll of the Minor Prophets, discovered in Wadi Murabbaʿat, are virtually identical – even in orthographic detail – to the Bible texts preserved by the tradition. They testify vividly to both the antiquity of the massoretic text of the Bible (that regularly used by Jews) and the accuracy of its transmission.

Other scrolls vary significantly from the biblical text as we know it. One Isaiah scroll from Qumran contains a "modernized" version of the book, in which the language has been simplified and the spelling altered to aid in reading and understanding the difficult Hebrew poetry. The Psalms scroll from Qumran contains non-biblical texts, which suggests either that it was not intended to be a copy of the biblical book at all (perhaps it functioned as a prayer book) or that the biblical canon of that locale differed from others with which we are familiar.

The discovery of a Hebrew copy of Jeremiah, substantially thinner than that preserved in the Hebrew Bible but sharing many features with the ancient Greek translation, has confirmed the suspicion that at least two Hebrew versions of the book existed in ancient times. Other newly discovered texts that differ from the traditional readings but resemble known ancient translations further demonstrate the early use of a range of biblical texts.

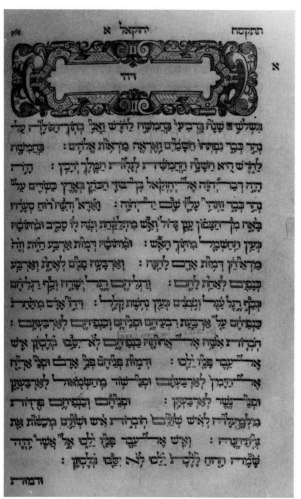

The first chapter of Ezekiel, containing the prophet's vision of the divine chariot in a whirlwind. The roots of the Hebrews letters appear in solid type, while supplementary letters are in outline form. Root letters that do not appear in the inflected form are added above the word. Jewish Public Library, Montreal
(Cat. no. 6)

Frontispiece from the Polyglot Bible of 1596, including illustrations of the Tower of Babel, four of Noah's descendants teaching their followers, and scenes from the second chapter of Acts. Jewish Public Library, Montreal
(Cat. no. 6)

Masada

Rome's general control of the Mediterranean world in the first century extended to Jerusalem and the surrounding territory, but as Jewish rebelliousness surpassed the limits of Roman tolerance, the country was systematically and viciously conquered. Masada was the final fortress to fall to the Romans, three years after the revolt that saw the Jerusalem Temple go up in flames in 70 CE. The lives and deaths of those who perished on this desert mountaintop are vividly described by Josephus, a Jewish historian of the first century whose pro-Roman presentation of the events has served as the basis for almost all attempts at their reconstruction.

The reports on those who died on Masada suggest interest in Scripture and Jewish religious practice, so the discovery of Bible fragments atop the mountain should be no surprise. But the complete destruction these people suffered seemed to discount any chance of recovering their books or writings, and the find of over a dozen scrolls, including pieces of Leviticus, Deuteronomy, Ezekiel, and Psalms, is thus all the more intriguing and noteworthy.

The text of Psalms recovered at Masada contains parts of chapters 81 to 85, the major parts being Psalms 82 and 83, which appear in the two middle columns of the illustration. Generally, the verses are divided into two stichs, and the manuscript follows this division by placing each in a different column of text.

The spelling is very close to that of the massoretic text. Occasional variations occur, as in the exchange of the suffix -*kh* in the scroll for -*k* in the Bible (81:8); the spelling of '*elohim* as '*lwhm* (82:6); omission of a conjunction, ʿ*mwn* for *w*ʿ*mwn* (83:8); substitution of '*lhym* for '*lhy* (83:14); and a slightly different division of the poetic stichs (83:10). These changes are all minor, and are typical of the types of variations that appear in medieval Bible manuscripts and early printed editions. In short, the portions of Psalms preserved in this scroll are virtually identical to those found in the massoretic text.

Like Job 38, Psalm 83 refers to a whirlwind. Most of this chapter entreats God to punish those who rebel against Him and requests that a series of natural calamities befall those nations worthy of feeling His wrath. The prayer includes requests that they be blown away in the wind, consumed in fire, and routed in a whirlwind, the same type of storm from which God revealed himself to Job. From these passages we see the strength of the biblical notion that the whirlwind represents God's presence and is one of the vehicles through which He acts in the world, in this case to mete out punishments.

Bible. Psalms. Hebrew
תהילים עם מעמדות(?)
[Tehillim ʿim Maʿamadot (?)]
Unidentified
B. Barry Levy, Montreal

5

Many of the Psalms were sung in the Jerusalem Temple, often with musical accompaniment, and the book later served as the foundation of both the Jewish and the Christian liturgies. Prayers were added freely, so that the Hebrew prayer book is no longer identical to the Psalter, but many sections (*Pesuqei de Zimra, Qabbalat Shabbat,* and *Hallel*) are taken almost exclusively from it, and its impact on the rest of the prayerbook is pervasive.

The spiritual attraction of the 150 chapters of the Psalter was so great that excerpts were frequently chanted in addition to the three daily prayer services, often on a weekly basis that required a daily recitation of approximately twenty chapters. Many special editions of the Psalms were published for this purpose, and this text, printed on parchment, was a particularly valuable one. Where defective, the volume was augmented with handwritten pages. The edges of many pages have been damaged where the writing has been rubbed off from excessive use by fervent supplicants.

Bible. Hebrew
Opus Quadripartitum Sacrae Scripturae, Continens S. Biblia siva Libros Veteris et Novi Testamenti Omnes Quadruplici Lingua Hebraica Graeca Latina et Germanica, Volume 1
Hamburg: 1596
Jewish Public Library, Montreal

6

The whirlwind was important for Job, but Elijah's experiences with it offer two

15 / God and Nature in Ancient Times

complementary treatments. At one point early in his life, while actively seeking God's presence, Elijah observed that God was not to be found in wind, nor in fire, nor in loud noise – all manifestations of a whirlwind or severe storm – but in a "soft, low voice" (*I Kings* 19:11-12). It is therefore interesting to discover that, at the conclusion of his earthly service, this same Elijah was taken to heaven in a chariot of fire, propelled by a whirlwind (*II Kings* 2). God may not have been limited to storms, or even been best symbolized by overwhelming natural forces, but He worked through them and used them to do His bidding.

This book is part of polyglot Bible, but unlike those described below, all of the texts in the different languages are not printed together on the same page. Unlike most Bibles, the roots of the Hebrew words are printed in solid letters, while added letters are hollow.

Francisco Ximenes De Cisneros of Toledo (1436–1517)
Biblia Sacra Polyglotta ...
[Complutensian Polyglot]
Alcala: 1514-17
Presbyterian College, Montreal

| 7 |

Brian Walton of London (ca. 1600–61)
Biblia Sacra Polyglotta ...
[London Polyglot Bible]
London: 1655-57
Presbyterian College, Montreal

| 8 |

The beginning of the prophetic book of Ezekiel contains the most detailed description of God's presence found in the Bible and, like several others, associates His appearance with a stormy north wind. This highly enigmatic passage describes the heavenly beings that move God's chariot and the human form of its rider, but carefully avoids suggesting that the rider is God himself by noting that it is merely "the appearance of the presence of the glory of the Lord."

The Bible has been translated into many ancient languages and virtually every extant modern one. In the fifteenth and subsequent centuries, its message was carried to many new regions of the world, even while European scholars clarified the accuracy of its text through comparison with the ancient versions (the Samaritan Pentateuch, the Greek and Aramaic translations, and other secondary translations made from them). These interests, coupled with the development of the technology needed for printing different scripts on a single page, contributed to the publication of polyglot Bibles that included both Jewish and Christian Scriptures.

The Complutensian Polyglot, prepared under the direction of Cardinal Ximenes, Archbishop of Toledo, is not uniform throughout because several books were unavailable in some languages, and all translations were not equally valued. The Hebrew text of the Pentateuch is accompanied by the Aramaic translation of Onkelos, as well as a Latin version of the Aramaic text, the Septuagint in Greek with an interlinear Latin translation, and the Vulgate in Latin. Subsequent volumes omitted the Aramaic translation and contained no Latin or Hebrew texts for those books not available in those languages (e.g., some apocryphal books). The important text-critical work of the editors and the beautiful type created by the printers render this volume very valuable from both scholarly and artistic standpoints.

The Walton Polyglot, though somewhat less attractive, is even more impressive. In addition to the Hebrew text and the Samaritan text of the Torah, it contains translations into Aramaic, Greek, Latin, Ethiopic, Arabic, Syriac and Persian, all in their native scripts and accompanied by Latin translations.

Bible. Pentateuch. Samaritan

תורה

Israel: Jacob ben Sadaqa, 1825
McGill University, Montreal

| 9 |

Many of the characters in the early genealogies of Genesis are relatively unknown, but Enoch, who is mentioned only in Gen. 5:18–24, later became the subject of much speculation. Enoch was the father of Methusaleh, who, at 969 years, set the biblical record for longevity. However, unlike the other people mentioned in chapter 5, whose deaths are carefully reported, of Enoch's we are told only that he "walked with God and is not, because God took him." This deliberately ambiguous description was later seen as a prototype for the ascension of Elijah, and, coupled with Enoch's being the father of the oldest man on record, stimulated great interest in his life story. Moreover, Enoch is reported to have lived for 365 years, and the cosmic associations were too obvious to escape later embellishment.

In Hellenistic times, Enoch became one of the most popular biblical characters and the subject of many independent books, several of which are found in the apocryphal literature. Because he was taken by God and his death was not mentioned, many believed that he was transported to heaven, where he observed the workings of the universe; many passages that report his observations served as a popular vehicle for the revelation of ancient meteorological, geographical, and astronomical speculations. Enoch's early post-biblical role as the revealer of scientific knowledge was lost in the Babylonian *Talmud, which never mentions him, probably because of his post-biblical associations with heterodox theological speculation. He did re-emerge, however, identified as Metatron, a leading heavenly figure in later kabbalistic literature.

The Samaritans are the oldest non-rabbinic Jewish sect in existence. Mentioned in the biblical book of Ezra, they were rejected by the Jewish returnees from the Babylonian exile, but were, according to their own reckoning, heirs to the authentic Biblical heritage. They also figure prominently in a number of passages in the New Testament and the Talmud. Unlike the rabbinic Jews, the Christians, and the Karaites, the Samaritans do not accept the sanctity of the entire Bible, and their scriptures are limited to the Pentateuch. Also, they copy their Pentateuchs in a script that resembles the paleo-Hebrew alphabet found in inscriptions and documents from biblical times, not that used by rabbinic scribes. Their version of the Pentateuch contains thousands of differences from the traditional one. Most are minor or of only linguistic significance, but occasionally they contain important literary, historical, or legal variants.

The beginning of chapter 7 of Spinoza's *Tractatus Theologico-Politicus*, which deals with Bible interpretation. Jewish Public Library, Montreal (Cat. no. 69)

Title page from the Offenbach, 1720, edition of Abraham bar Hiyya's *Tzurat HaAretz*. National Library of Canada, Jacob M. Lowy Collection (Cat. no. 13)

Chapter Two

astronomy

איפה היית ביסדי ארץ
הגד אם ידעת בינה!

איוב לח, ד

The heavens declare the glory of the Lord ...

Psalm 19:1

Said the Holy One, Blessed Be He ... "I created twelve stellar regions in the sky, and in each one I created thirty divisions; and in each division, I created thirty legions; and in each legion, I created thirty units; and in each unit, I created thirty cohorts; and in each cohort, I created thirty camps; and in each camp, I placed 3,650,000,000 stars.

Babylonian Talmud, Berakhot 32b

[Forgive me, O Lord, for the sin of] looking too much at the rainbow and the moon.

Rabbi Hayyim Joseph David Azulai, "Personal Confession"

The movement of shadows in relation to the movement of the sun. Abraham bar Hiyya, *Tzurat HaAretz,* Offenbach, 1720. National Library of Canada, Jacob M. Lowy Collection (Cat. no. 13)

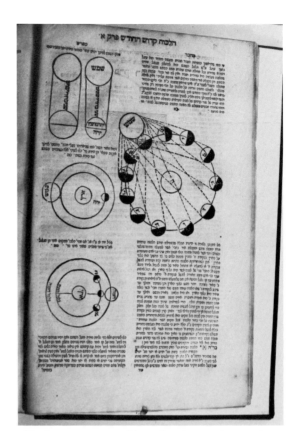

The changing positions of the earth and moon relative to the sun that cause the changing phases of the moon. Jonathan ben Joseph of Ruzhany, *Yeshuʿah BeYisraʾel,* Frankfort on Main, 1720. National Library of Canada, Jacob M. Lowy Collection (Cat. no. 136)

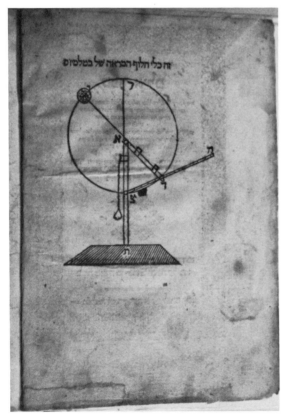

A reconstruction of a piece of astronomical equipment used by Ptolemy. Delmedigo, *Sefer Elim – Maʿayan Ganim.* National Library of Canada, Jacob M. Lowy Collection (Cat. no. 72)

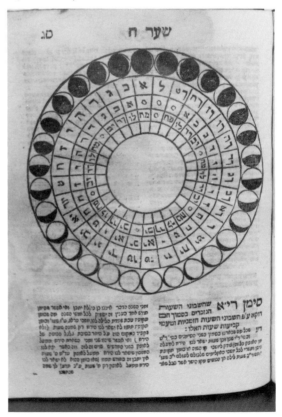

Phases of the moon seen as a function of the relative positioning of the earth, the moon and the sun. Tobias Cohn, *Maʿaseh Tuviah,* Jesnitz, 1721. Jewish Public Library, Montreal (Cat. no. 74)

Claudius Ptolemaeus of Alexandria (second century)

ספר אלמגסטי לבטלמיוס
[Sefer Almagesti]
Sixteenth century
Bibliothèque Nationale, Paris (Heb. MS no. 1019)

Hellenistic Egypt was a centre of scientific learning. It was home to, among others, Ptolemy, Archimedes, and *Euclid; it housed the best library in the world; and it provided excellent opportunities for scholars to study, organize, and assimilate the vast scientific legacies of the Greco-Roman and Mesopotamian civilizations. A tradition recorded by Josephus (*Antiquities I*, 167) and by several other writers credited Abraham with teaching Eastern mathematics and astronomy to the Egyptians. This notion has little to commend it as history, but it does reflect the fact that much Eastern scientific and mathematical knowledge was available in Egypt and that Jews there identified with it.

Ptolemy, about whom very little is known, wrote on optics, geography, mathematics and philosophy, but his most famous composition is the Greek work, entitled *Mathematike Suntaxis*, known through the Westernized form of its Arabic name, *Almagest*. The *Almagest* was translated into Syriac, Arabic, Hebrew (by Jacob Anatoli), Latin (from the Arabic, by Gerard of Cremona), and other languages, and was the single most important astronomical work composed before the scientific revolution of the sixteenth and seventeenth centuries. Accordingly, it played a direct role in virtually every medieval and renaissance Jewish book on astronomy. In fact, after Ptolemy's theories were replaced, some early modern writers engaged in extensive rationalizing to justify the religious significance that had been attributed to his teachings by medieval rabbis.

Ptolemy was aware that the stars and planets do not move around the earth in perfect circles. He accepted the principle of geocentricity, but he knew that more sophisticated models were needed to account for the observed irregularities in the stellar notions. His models employed mathematical devices known as epicycles, eccentrics, and equants, which were developed in part by Apollonius, the famous geometer. Both the models and the observational data associated with the Ptolemaic system were refined over the centuries; in fact, for well over a thousand years the Ptolemaic theory proved itself quite credible for naked-eye astronomers.

Abu'l-ʿAbbas Ahmad ben Muhammad ben Kathir Al-Farghani (Alfraganus) of Baghdad (ninth century)

קיצור אלמגסטי
[Kitzur Almagesti]
Spanish, square script, fourteenth century
Hebrew Union College, Cincinnati (MS no. 891.3)

Al-Farghani served as an engineer in Egypt and is best known for his very popular astronomical work. He wrote several tractates on the astrolabe, but his most famous composition, known under various titles, was an abridgment of *Ptolemy's *Almagest*. Ptolemy's work was presented in thirteen large books; Al-Farghani regrouped it into thirty small chapters, omitted many details and mathematical discussions, and added his own material, especially in the first chapter. He wrote in Arabic, and the work was translated into Latin at least three times, once from the Hebrew translation by Jacob Anatoli.

Masha' Allah of Basra, Damascus (eighth–ninth century)
Liber Novem Judicum in Judiciis Astrorum
Venice: Petrus Liechtenstein, 1509
Hebrew Union College, Cincinnati

One of the oldest Arabic authorities on astrology, Mash' Allah (or Mash'allah), and three of his contemporaries were responsible for drawing the horoscope for the founding of Baghdad in 762. Mash' Allah, a Jew, served in the courts of several caliphs and had an appreciable impact on both Eastern and Western writers, especially through his work on astrolabes.

Ibn al-Nadim attributed nineteeen works to Masha' Allah (*Fihrist*, chapter 7). Most are devoted to astrological predictions, but others deal with astronomy, astrolabes, meteorology, and scientific matters that have some bearing on astrology. Perhaps his most

An eclipse. Abraham bar Hiyya, *Tzurat HaAretz*, Offenbach, 1720. National Library of Canada, Jacob M. Lowy Collection
(Cat. no. 13)

A boy falling into the water due to astral influences. Ibn Sahulah, *Meshal HaQadmoni*. Annenberg Research Institute, Philadelphia (Cat. no. 22)

An eclipse. Abraham bar Hiyya, *Tzurat HaAretz*, Offenbach, 1720. National Library of Canada, Jacob M. Lowy Collection
(Cat. no. 13)

Four arguments supporting the claim that the universe is geocentric, with a graphic clarification. Tobias Cohn, *Ma'aseh Tuviah*, Venice, 1708. National Library of Canada, Jacob M. Lowy Collection
(Cat. no. 73)

interesting work is the lost *On Conjunctions, Religions and Peoples,* an astrological history of mankind that has been recovered, in part, from citations in later authors. It is based on the Sasanian theory that significant religious and political events, for example, the flood, the nativity of Jesus, and the rise of Islam, are timed to the conjunctions of Jupiter and Saturn, at approximately twenty-year intervals.

Masha' Allah's works appear to have been preserved only in translation. Several of his books were translated into Hebrew by Abraham *Ibn Ezra and are found in the Bibliothèque Nationale, Hebrew manuscript no. 1051.

Abraham Bar Hiyya of Barcelona (ca. 1065–ca. 1143)

חכמת החזיון, ספר צורת הארץ
[*Hokhmat HaHizzayon, Part I*]
[*Sefer Tzurat HaAretz, Sphaera Mundi*]
Offenbach: Bonaventura de Naye, 1720
National Library of Canada, Ottawa (Jacob M. Lowy Collection)

| 13 |

חכמת החזיון, חשבון מהלכת הככבים
[*Hokhmat HaHizzayon, Part II*]
[*Heshbon Mehalekhot HaKokhavim*]
Syracuse on the Isle of Sicily: Sholom bar Shelomo HaYerushalmi, 28 Adar 5244 (1484)
The Vatican (MS Ebr. no. 379)

| 14 |

Abraham Bar Hiyya (or Hayya) lived much of his life in Barcelona, where he was a respected philosopher and scientist, and also served as a surveyor in France or Provence. He was the first philosopher to write in Hebrew (the efforts of his Jewish predecessors having been limited to Greek and Arabic) and the first to present Ptolemaic astronomy extensively in that language. Accordingly, he made important contributions to the Hebrew philosophical and scientific lexicon and aided subsequent generations of Jews, as well as contemporaries not living in Arabic-speaking countries, in learning astronomy. But unlike many later writers who wrote only in Hebrew, he was comfortable in other languages, and even contributed to the translation of scientific works from Arabic to Latin.

Bar Hiyya was a committed astrologer, and his well-known *Megillat HaMegalleh* – with sections on time, the age of the universe, resurrection, the soul, and the date of the messiah – contains major astrological discussions. Among the few known facts of his life is a disagreement with Judah ben Barzillai al-Bargeloni which resulted from Bar Hiyya's trying to have him postpone a wedding because the astrological timing was inappropriate. Al-Bargeloni equated astrology with idolatry, and refused.

Higayon HaNefesh HaAzuvah is Bar Hiyya's most famous treatise, and deals with cosmogony, cosmology, psychology, and ethics. His primary contribution, however, was in the physical sciences. His *Yesodei HaTevunah uMigdal HaEmunah* has been described as the first encyclopaedic work in Hebrew, and includes sections on geometry, mathematics, optics, and music. According to historians of mathematics, his *Hibbur HaMeshihah VeHaTishboret* was the first exposition of Arabic algebra written in Europe, the first European solution of the quadratic equation $x^2 - ax + b = 0$, and the first Hebrew work to define the area of a circle as the square of the radius multiplied by π.

Hokhmat HaHizzayon, Bar Hiyya's main astronomical work, contained two parts that have often been considered separate books. The first, *Tzurat HaAretz VeTavnit Kaddurei HaRaqia^c* (also known by a number of slightly different titles), is a geographical and astronomical text in ten sections. It includes discussion of the spherical shape of the world and detailed information about the earth's zones, the sun, the moon, eclipses, the planets, the stars, and the constellations. The second part, *Heshbon Mehalekhot HaKokhavim,* contains astronomical tables modeled on those of Al-Battani (a ninth-century Arabic astronomer), discussions of the stars' courses, and an analysis of the process by which the calendar is intercalated. The latter was also the subject of his *Sefer HaIbbur,* or *Sod HaIbbur,* written in 1122.

Vatican manuscript Ebr. no. 379 is a compendium of ten astronomical texts, also including *Sefer Orah Selulah,* by Isaac ben Solomon ben Alhadib, a general astronomical work in nine chapters; *Sha^{c}ar HaShamayim,* by Isaac *Israeli; and two sets of astronomical tables. As well, it holds five works on the production and use of astronomical equipment: *Iggerit HaIstrolab,* by Shalom ben Solomon; *Kelilat Yofi,* by Elijah HaKohen De Monte Alto; *Rova^c Yisra'el,* by Jacob ben Makhir Ibn Tibbon; *Keli Hemdah,* by Isaac ben Solomon; and *Bei'ur MiYermiah HaKohen Al Asiyat HaKadur.*

A page from Abraham Ibn Ezra's *Sefer HaItztrolab*. Bibliothèque Nationale, Paris, MS Heb. 1053 (Cat. no. 15)

The Basle edition of *Tzurat HaAretz* contains notes by Sabastian *Muenster (1489–1552). A student of Elijah Levita, Muenster was a philologist and astronomer of note who edited or wrote many Hebrew and scientific works, including Ptolemy's *Geographia*, the Hebrew Bible with the first Protestant translation from Hebrew into Latin and notes, a trilingual dictionary of Hebrew, Greek, and Latin, and the annotations to the Latin translation of *Tzurat HaAretz*. Muenster's most famous work is *Cosmographia*, which was published in over forty editions. His edition of *Tzurat HaAretz* is bound with *Qitzur Melekhet HaMispar* by Elijah *Mizrahi. The 1720 edition of *Tzurat HaHaretz* is bound with *Sacrobosco's *Sefer Asphaera*.

Abraham ben Meir Ibn Ezra (1089–1164)

ספר האאצטרולב
[Sefer HaIstrolab]
Sixteenth century
Bibliothèque Nationale, Paris (Heb. MS no. 1053)

15

ראשית הכמה
[Reishit Hokhmah]
Fourteenth century
Bibliothèque Nationale, Paris (Heb. MS no. 1045)

16

De Nativitatibus
Venice: E. Ratdolt, 1485
Hebrew Union College, Cincinnati

17

*Ibn Ezra was extermely versatile and prolific. He is best known for his Bible commentaries, which reflect a deep and abiding concern for the correct grammatical appreciation of the text, and for his essays on the Hebrew language. However, his contributions in the areas of science, philosophy, and poetry, though perhaps less well known, are no less noteworthy. Ibn Ezra's philological approach to the Bible helped spread the contributions of Arabic linguistic science to Jews who did not know Arabic, and this openness to the works of non-Jewish scientists appears in his other writings as well. His constant wandering during much of his adult life helped to popularize his books but also left them scattered and difficult to collect. Many compositions exist in multiple copies, some in more than one recension; others are lost.

Ibn Ezra was a devoted astrologer, and the popularity of his astrological writings is evidenced by the number of manuscript copies preserved, the fact that many of his works were translated into European languages, and the attribution to him of some astrological works written by others. Many of his books on astrology were composed during his stay in France. These include *Sefer HaMoladot* – in Latin, *De Nativitatibus* – *Reishit Hokhmah*, *Sefer HaTe'amim*, *Sefer HaMivharim*, *Sefer HaMe'orot*, and *Sefer HaOlam*, all of which are found in Paris Hebrew manuscript no. 1045 and some of which exist in two recensions. A number of his compositions deal with the astrolabe, which he believed to be similar in function to the biblical high priest's breastplate (*hoshen mishpat*) and useful in predicting the future. He integrated many aspects of science into his Bible commentaries (see, for example, his discussions of the difficulty of calculating precise times in Ex. 11:4 and his explanation for the difference in the speeds of light and sound in Ex. 20:1). He did the same with astrology, for which he was criticized by some later writers, particularly Samuel David Luzzatto.

Hebrew manuscript no. 1053 also contains three other works on astronomical equipment: Mordecai *Comtino's *Sefer Keli Nehoshet*, *Bi'ur Asiyat HaKadur*, and Qosta Ibn Luqa's *Sefer HaMa'aseh BeKadur HaGalgal*. Hebrew manuscript no. 1045 contains nine of Ibn Ezra's astrological works and his *Peirush Keli Nehoshet* on the astrolabe; astronomical charts of Abraham *Bar Hiyya; two works by *Mashallah; *Sefer Hermes*, an astrological work translated from Arabic; and several other items. Hebrew manuscript no. 1051 (cat. no. 51) contains Ibn Ezra's *Sefer HaMoladot* and *Sefer HaShe'eilot*, as well as his work on division and five books on astronomy and astrology by other writers.

Johannes Sacrobosco (died 1256)

ספר מראה האופנים
[Sefer Mareh HaOfanim] [Sphaera mundi]
German cursive, Tevet 5186 (1426)
Montefiore Endowment, Jews College, London (MS no. 424)

18

Sphaera mundi, which incorporates elements from both Ptolemy's *Almagest and *Al-Farghani's "abridgment" of it, became one of the best-known medieval works on astronomy. It was translated into Hebrew by Solomon ben Abraham Avigdor in 1399, and was annotated by a number of writers, in both Latin and Hebrew.

Isaac ibn Said (Sid) of Toledo (thirteenth century)

לוחות מהמלך דון אלפונסו
[Luhot MeHaMelekh Don Alfonso]
Fifteenth century
The Vatican (Ebr. no. 382)

19

In the thirteenth century, many Jews were deeply involved in the astronomical studies at the court of Alfonso X "the Wise." Although Alfonso's treatment of the Jews deteriorated during his reign and became quite hostile at times, their contributions to his scientific activities, including the composition of the "Alfonsine Tables," drawn in 1252–56 under the direction of Isaac Ibn Said (the synagogue cantor) and Judah ben Moses HaCohen, were considerable. The Alfonsine Tables were based, in part, on the Toledo Tables, composed in the previous century by twelve astronomers (including a number of Jews) under the direction of Al-Zarkali, and were translated into a number of languages and used well into the seventeenth century.

Vatican manuscript Ebr. no. 382 contains the Alfonsine Tables translated into Hebrew; a Hebrew commentary on the tables; a translation of *Sacrobosco's *Sphaera mundi*, *Mareh HaOfanim*; Israel ben Saniel's *Mishpetei HaKokhavim*; and a commentary on the pseudo-Aristotelian *Sefer HaPeri (Liber Fructus)*.

Other astronomical tables were compiled by Abraham *Bar Hiyya, Abraham *Ibn Ezra, Isaac *Israeli, *Gersonides, Jacob *Poel, and Abraham *Zacuto.

Jacob ben David Bonet or Poel (fourteenth century)

לוחות
Astronomical Tables
Bologna: Italian script, 1395
Biblioteca Palatina, Parma (MS no. 2275)

20

Son of a Spanish physician and astronomer who manufactured astronomical equipment for Pedro IV of Aragon, Jacob is best known for his astronomical tables. These were prepared for Perpignan for the year 1361, were translated into Latin in the fifteenth century, and were the subject of many Hebrew commentaries.

Mordecai (Angelo) ben Abraham Finzi of Mantua (fifteenth century)

לוחות
Astronomical Tables
Italian cursive
Montefiore Endowment, Jews College, London (MS no. 428)

21

Finzi, an important fifteenth-century Italian Jewish scientist, wrote on Hebrew grammar and mathematics and translated several Arabic mathematical works into Hebrew. His most famous work is a collection of astronomical tables, which was published in Latin around 1479, but it represents only a small part of his contribution.

Isaac ben Solomon ibn abi Sahulah of Guadalajara (late thirteenth century)

משל הקדמוני

[Meshal HaQadmoni]
Frankfort on the Oder: Johann Christopher Beckman, 1745 (?)
Annenberg Research Institute, Philadelphia

22

Sahulah lived in the same city as Moses De Leon, author of the *Zohar*, and in all likelihood knew him; he was certainly acquainted with his work. Sahulah is reported to have written commentaries on several books of the Hagiographa, but his reputation is based on *Meshal HaQadmoni*, probably the most popular work of Hebrew fables ever composed.

Meshal HaQadmoni is a collection of animal fables, written in *maqama* style, for teaching morals. The work is divided into five sections that deal with the intellect, penitence, correct advice, humility, and fear of God, and is reminiscent of Aesop's fables, both in content and in manner of illustration. Unlike the characters in Aesop's stories, who are unlearned even when wise, those in *Meshal HaQadmoni* frequently demonstrate an acquaintance with rabbinic literature, the Zohar, and other Jewish and scientific subjects.

Isaac ben Joseph Israeli of Toledo (early fourteenth century)

יסוד עולם

[Yesod Olam]
Spanish script, fourteenth or fifteenth century
Biblioteca Palatina, Parma (MS no. 3165)

23

Berlin: 1777
Jewish Public Library, Montreal

24

Isaac Israeli, a contemporary of Asher ben Yehiel, the famous talmudic commentator and halakhic authority, composed *Yesod Olam* for this esteemed rabbi and his students and children, presumably including his son *Jacob ben Asher. Completed in 1310, it deals with the range of mathematical and astronomical issues that emerge in the halakhic literature, particularly in the Talmud and in *Maimonides' *Hilkhot Qiddush HaHodesh*.

Israeli outlined his goal as an analysis of the principles and manner of calculating rules of the calendar, especially the intercalation of the extra month at the end of a leap year; but his effort is much more extensive than many of the later works devoted to this topic. He began, in Book *I*, by defining "point," "angle," and other geometric terms, and moved on to the study of triangles and spherical trigonometry. Only after establishing this foundation did he turn to astronomy.

Books *II* and *III* deal with the heavens and earth, including the creation of the universe, the movements of the celestial bodies, and the signs of the zodiac. Also discussed are the measurement of dry land, the nature of sunrise and sunset, calculation of periods of time, and especially the heavenly spheres. Book *IV* is devoted to the moon, the lunar year, the nineteen-year calendrical cycle, determining the new moon through both observation and calculation, and related talmudic teachings. It concludes, in chapter 18, with a summary of world history that draws upon Abraham ben David's *Sefer HaQabbalah*. The fifth and final book is the most complex, as it applies the theories discussed earlier to the composition of detailed charts and tables about the sun, the moon, the months, determination of Rosh Hodesh, and related matters. As is true of so many other medieval scientific works, attention is also given to related religious issues, such as the nature of biblical miracles.

The 1777 edition was edited by *Barukh of Shklov, who supposedly added the sketches, but a careful comparison with the manuscript tradition is needed to determine the precise nature of his contributions, because the manuscripts themselves are full of drawings.

The Cracow 1580–81 edition of Abraham *Zacuto's *Sefer Yuhasin* contains a small section of notes by Moses *Isserles on the historical part of Israeli's *Yesod Olam*. Isserles' treatment of astronomy (the major concern of *Yesod Olam*) appears in his *Torat HaOlah* and in an unpublished commentary on *Mehalekh HaKokhavim*, Ephraim Mizrahi's translation of Peurbach's *Theorica Planetorum*. This does not qualify as an independent astronomical work, but clearly indicates the extent to which Isserles studied astronomy. His comments on *Yesod Olam* show the importance he attributed to the work, even if most of his attention was directed to augmenting and correcting its historical section.

Gersonides, *De instrumento secretorum revelatore,* Bibliothèque Nationale, Paris, MS Lat. 7293 (Cat. no. 26)

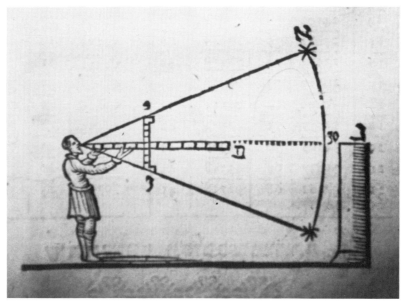

Use of the Jacob's Staff, reportedly invented by Gersonides. Delmedigo, *Sefer Elim–Ma'ayan Ganim.* National Library of Canada, Jacob M. Lowy Collection (Cat. no. 72)

Levi ben Gershon of Orange (1277-1344)

מלחמות ה'

[Milhamot Adonai]
Abraham Saul, 14 Tevet 5270 (1510)
Bibliothèque Nationale, Paris (Heb. no. 725)

De instrumento secretorum revelatore
[Milhamot Adonai]
Fourteenth century
Bibliothèque Nationale, Paris (Lat. no. 7923)

Levi is known in different languages as *Gersonides, Ralbag, Maestre Leo de Bagnols, and Magister Leo Hebreus, and the details of his life are relatively obscure. Many scholars identified Gershon, his father, as Gershon ben Solomon of Arles, author of the scientific encyclopaedia *Sha'ar HaShamayim*, but this association has little to support it and has generally been abandoned. Few medieval thinkers matched Gersonides' impressive literary output of commentaries and original essays, most of which integrate religious and scientific concerns.

A respected religious leader, Gersonides wrote a number of liturgical pieces and was consulted on several important halakhic decisions, but the formerly accepted attribution to him of commentaries on parts of two talmudic tractates is questionable. Even so, his work demonstrates mastery of the *Talmud, and his contributions to science and Bible interpretation are well documented and very important.

Gersonides' commentaries on the Torah, the Former Prophets, and nine books of the Hagiographa, including Job, were among the first commentaries printed. Discussions of philosophy and science pervade his interpretations, and his naturalistic presentations of the Bible's miracle stories were often discussed by later writers. Moreover, his exegetical activities were not limited to the Bible; he composed commentaries of a philosophic-scientific nature on numerous compositions by *Averroes (often based on Aristotle). His independent books were devoted to philosophy, science, and mathematics.

Gersonides' most important philosophic-scientific work, *Sefer Milhamot Adonai*, deals with many of the classic issues of medieval Jewish philosophy, and includes a full treatise on astronomy that was omitted from most copies and was ultimately treated as in independent composition. It contains sections on trigonometry; the Jacob's Staff; the motion of heavenly bodies; *Ptolemy's model of lunar motion and Gersonides' improvements; solar and planetary motion; and the order, sizes, and distances of the planets and fixed stars.

Gersonides claimed credit for inventing the observational tool known as the Jacob's Staff, which simplified calculating angular distances. The staff consists of a rod about five feet long and one or more sliding plates. The stars to be measured are sighted at the ends of the plate, after which the angle can be calculated. Gersonides also correctly located the precise centre of vision in the eye and was able to compensate for distortion due to placement of the eye itself. His resultant measurements were therefore both simple and accurate.

According to B.R. Goldstein, to whom we are indebted for much of what is known of his astronomical work, Gersonides made a number of important modifications to Ptolemy's theories and claimed that actual observations did not confirm the theory of epicycles and eccentrics. He also developed a new theory of lunar motion that corrected a number of errors in Ptolemy's theory, and he discussed and criticized Al-Bitruji (Alpetragius), a twelfth-century Spanish Moslem whose astronomical works had been translated by Moses ibn Tibbon. Despite these advances, he was convinced of the value of astrological speculation and rejected Maimonides' radical denunciation of it.

Gersonides emphasized the need for observation of celestial phenomena to complement astronomical theory, and his own observations often underlay his critiques of his predecessors. Opportunities to observe unusual phenomena were not overlooked, even when religious obligations might have provided excuses. He reported observing a lunar eclipse on the first night of Sukkot in 1335, in the company of a distinguished Christian who recorded his comments.

The importance of his astronomical work and its popularity among Christian scholars is illustrated by the extent to which it was translated into Latin, motivated in part by the desire of Pope Clement VI to consult it. The Hebrew text, however, was never fully published.

שש כנפים

אנה היה יוצאין יוסיפהם	והמאחז ועקות אבהם	ענן דין יחסבן קל וקטו	טלטטרך ולא יעדך רבהס
בעמיה וענינה ימיתהם	ובניהם וקדרות החודים	דיהיסהו יחסיהו לחלון	בקדוק רב וכנפוה קלפד
הנה כטב כבעב	שהם טהעשף מהיהם	והסה עי טעב בטפר	לנעשה חקוק ומויהם
ויוט לטסיות התהבטב	ביעה הענבטו טקוהב	ויוטו על ננ שלטר	וינטו ולא יטע עיד
וידך זו להושפ בה וקד	ודיהט מטטה ועוויב	וקשר תלא יעל אנהב	להחא יקרה עוב כטב קשה

אמר מנואל בן יעקב

כטל הכנבים אם חויה יהב ומטיה ימלאטי על העפות יוטל דיטים לטר עיטן יברוץ והמאטי לטטב החולם
והכנבים החמישים וקלרהות הירחים הסטוים דין קלוקט יעד וכדקוק רב אחד לו בן טטיה סטורץ
המחטבה יריח ושמרות ואמרתי כמוה היה אחריה כטלאהן בנטב ואטב קטול הרב הוא כמו פלו טעדן בורץ
חטבן לחטבל פעקני הייתורך לא מה יוהבנה ולא מרימטב ולא שך החילפו ולא חימן המיהה לא מורץ ולא
בריח כן עו הטל יוחטן בטלח יב עו עטטעמו יוטל בטה לטעבא ומיל יקהבן הוטל יוחל יודר מאד חטטעונב
הראות וחמטר יד עטר החטבבן יוח ומטיהט יעוב היאחד עטב טעב המונהב אלא אחר החטבטב לעוטעך החילפו
יומן החילב על הטעב קטרית בטול חטבבן וזה לעם עטב ביטה בטחוהייני לוהרט ומו וטברוטי המדי יומרי ל
וקרה לטטים יטטה חילפו יודה ביטטטוט בן הטעב ונה החטטן המדיקן מו טלטו ול תלקם טהאב סבל טך הילעל יעד
לטעב לטטען הטע עבד טעב פטואב אולא עטר חטטעט לטיוך על דעב הסט טבטן וזה רחל לש פלאות בטלאטית
אק רעי לחמוך הלעה טל פט עטב האו והה מטכל הול לטיט לטטנב בכל החטבטוץ בן טעלל יחלא ובה קן פרוקה
טבה חחיע וטיקון והה דבה מטבעטנב כומפטא חאפט את וה עה חוטמפטו הך החילעל הטעב שלטעות לטעב טלטים
היומטין והעני לב דעט בן כטפו בעה תעוטו וכסו הטטבך הך החילעל וקרה לטטעים טיטה החילעל יחה העטנ
בן חטעוב וגן חטעבן הוה החעדקדק קעל כן טעב יטלעטי עבד טטא יוחטנב אלא אחר החטעבנב אחר קטטע טך
עב חיה ובוה וטו לטלך חחטטבן החוי יל חטורך על שהוריות ירח טטעה חית כטת תחה טעקרך על חטולך
יד טע ם עם קך לנבה אטור פרט טטהחטטן חו העבר חוהמול עוה חטת הקטריך יטור פרט זה לן דין פ רחו
להט לקור של מיהך היי טעב שם קן לבט הטעבן טטאטבר יטלף לטעון זחון יטיר אם החולך חוהעטר הטעה
עיהטלך תטוף פרט וגן החול יחו העטבר דתמיתי וקן ליפע יטם חוטטבבט וזה ויקרך לטטעים עישה החילעל
מוה קטעה בן חטטטב וגן הא החטטטן האורחקיד לו טעב קן טעב וטטוט יעך כל קטרטיט הטטיט
יקרה לטטעים עיטהב החילעל פן חטעטב וגן זה חטטבן המדיקין והטטבן לו טעב וכטע וטטעו הטעטמטם טחטעט
יהיה זה החטטן יוגר יומדקד וידע דן וטען י שהמטץ בדיק יהטי  יקרהו אלא וילאיו
להטעיד להוטטטב החטבן זה נקעט שטטטב בטע ודה יהטיפטב החולמה האהן וקדוק
וקן הונש בחנון עם בחות לקרה מול חו טעבר חורית או בקריות יהיה אונטו טעטר הטוות הקן וקליה
לטטם ל בטטב עלהדין יהפטר אז עטבן שך החילעל ועבן ידעני אים טעבן חקרה של טיהל היה טעה וב
יטלאטי חטטטף עבד והחילך טמטע על זה החטטבן הקלוהקטו חו תבעה מלאטטב וקטבעה הלא לעורות ואם ע על
בעה ראיות מטהבות חימוט ייוקרקן חטטפ ויטהירוהו יך העטטה ואין לטו עיבטע חחילף רחב פטיות והברית מה עט לטחור
קריאת כנבים טעב שהוא לטטיוג בעלעבה אטור עך קטע נוחנבה על עי יחא לחטטנב עגם לאט הריך לטע וחטבן טלטו
לטוטפים

Immanuel ben Jacob Bonfils of Tarascon (fourteenth century)

שש כנפים
[Shesh Kenafayim]
Syrian, rabbinic script
Montefiore Endowment, Jews College, London (MS no. 423)

Immanuel, a younger contemporary of *Gersonides, was an important astronomer and mathematician in Orange and Tarascon (Provence). A prolific author, he composed numerous mathematical and astronomical works.

His most famous book is *Shesh Kenafayim*, a six-part collection of astronomical charts which deals with conjunctions, eclipses, and similar phenomena. The charts, influenced by Gersonides but based, in large measure, on the work of Al-Batani (a ninth-century Arabic astronomer whose work was used by *Maimonides but rejected by Gersonides), are organized according to the Hebrew calendar and modified to apply to France. *Shesh Kenafayim* was translated into both Latin and Greek in the early fifteenth century and was used well past that time.

Montefiore manuscript no. 423 contains a number of important astronomical works: parts of Abraham *Bar Hiyya's *Tzurat HaAretz; a treatise on the astrolabe according to *Ptolemy; the translation of Peurbach's *Theorica Planetorum*, a popular work commented on by Moses *Isserles and Moses *Almosnino; and an incomplete copy of Abraham *Ibn Ezra's *Keli Nehoshet on the astrolabe.

Anthology of Astrological Texts
Spanish, rabbinic script
Montefiore Endowment, Jews College, London (MS no. 425)

Astrolabes are astronomical tools, developed by the ancient Greeks and improved by the medieval Moslems and Jews, for determining the locations of various celestial bodies and for navigation. Popular belief in the influence of the stars on human events encouraged development of astrolabes for astrological purposes, to the point where this use superceded the first one, but only in the eighteenth century was the astrolabe abandoned for the quadrant. Contributors to the development of the astrolabe and the literature about it include *Mashallah, Abraham *Ibn Ezra, and Abraham *Zacuto.

Montefiore manuscript no. 425, written on both parchment and paper, contains three treatises on the astrolabe: an anonymous set of instructions for preparing one, a forty-chapter essay translated from Arabic by Jacob ben Makhir, and a treatise by Judah ben Eliezer. In addition, it contains a brief note on the stars of the zodiac by *Gersonides, various astrological and meteorological rules, astronomical charts, lists of the latitudes and longitudes of cities in Asia, Africa, and Europe, and other items. It concludes with a diagram of an astrolabe.

Profiat Duran of Catalonia (died ca. 1414)

מעשה חושב
[Maʿaseh Hoshev]
Rabbinic script, fifteenth century
Biblioteca Palatina, Parma (MS no. 2776)

Isaac ben Moses HaLevi, also known under his baptised name, Honoratus de Bonafide, is best recognized as Prophiat Duran or Efodi. Forcibly converted, he was an important anti-Christian polemicist, whose works include *Al Tehi KaAvotekha* and *Kelimat HaGoyim*. His most famous work is *Maʿaseh Efod*, a significant book on the Hebrew language that contains chapters on philosophical and linguistic matters which are much more extensive than those normally found in grammatical treatises.

A many-faceted author, Duran also composed notes on parts of *II Samuel* on some of Abraham Ibn Ezra's poems and parts of his Torah commentary and on *Averroes' treatise on mathematical astronomy. His commentary on *Maimonides' *Guide for the Perplexed* is well known, and a number of his letters and brief communications have been published.

Maʿaseh Hoshev, in twenty-nine chapters, deals with astronomy and the calendar. Chapter 23, on new moons and calendrical intercalations, is written in poetry. The work is cited a number of times in *Kol Yehudah*, Judah Moscato's commentary on the *Kuzari.

Tobias Cohn's presentation of the solar system. Note the discussions of the theories of Tycho Brahe and Copernicus, with which the pages begin. *Ma'aseh Tuviah*, Venice, 1708. National Library of Canada, Jacob M. Lowy Collection (Cat. no. 73)

An excerpt from Abraham Zacuto's tables. Montefiore Endowment, Jews College, London, MS #426 (Cat. no. 37)

Abraham ben Samuel Zacuto of Spain, Portugal, and Jerusalem 1452–ca. 1515)

ספר יוחסין
[Sefer Yuhasin]
Cracow: 1580–81
National Library of Canada, Ottawa (Jacob M. Lowy Collection)

30

Almanach Perpetuum
Venice: Petru Liechtenstein, 1502
Columbia University, New York

31

Abraham Zacuto studied astronomy at the university in Salamanca and became one of the leading astronomical consultants to the Spanish and Portuguese explorers. He was personally involved in the travel preparations of Vasco de Gama, and Columbus made extensive use of his charts. Zacuto is credited with constructing the first copper astrolabe; his astronomical tables, which were sought by many sea captains, improved upon the *Alphonsine Tables, the best in existence until then. It is reported that Columbus used Zacuto's prediction of an eclipse to intimidate the natives he met, a story popularized in Mark Twain's *A Connecticut Yankee in King Arthur's Court*.

In addition to his many astronomical contributions, Zacuto composed *Sefer Yuhasin*, an important historical work first published in Constantinople in 1566. Part VI, which contains an outline of the history of the non-Jewish world, includes much information on the history of science.

HaHibbur HaGadol on astronomy (which also circulates under at least a half-dozen other titles) was written at the request of Zacuto's patron, the bishop of Salamanca, and has enjoyed tremendous popularity. It was translated into Spanish, and an abridgment was translated into both Latin and Spanish by Joseph Vicinho, one of Zacuto's students. An Arabic translation was also made, the Spanish version was transliterated into Hebrew, and a revised Latin translation was published in 1496.

Nicolaus Copernicus of Frauenberg, Poland (1473-1543)
Nicolai Copernici de Revolutionibus Orbium coelestium, Libri VI.
Norimberg: Ioh. Petreium, 1543
McGill Univeristy, Montreal (Osler Library, no. 566)

32

Copernicus, a physician and canon, made his most famous contribution to the history of astronomy by refuting *Ptolemy's teachings that the sun and planets revolve around the earth and popularizing the long-known but often-ignored heliocentric alternative. The mathematical confirmation of his proposal was later provided by Galileo and Kepler.

Copernicus' writings were criticized by some Jewish scientists (e.g., Tobias *Cohn) and supported by others (e.g., David *Gans). They became symbolic of the new scientific freedom that emerged in the sixteenth and subsequent centuries and helped widen the gap between religion and science.

This copy of the *editio princeps* is particularly interesting because of the Hebrew inscription found on the title page. The brief text articulates a strong commitment to God, denoted by several of His biblical names. In all likelihood, it was added by a Christian who, in response to recently developed Christian interest in Hebrew, had mastered some elements of the language. One can only speculate if the Hebrew credo was intended as a denial of Copernicus' teachings, which were sometimes challenged on theological grounds by both Jews and Christians.

Moses ben Barukh Almosnino of Salonika (ca. 1515–ca. 1580)

פרוש כדור העולם ושער השמים
[Peirush Kadur HaOlam and Sha'ar HaShamayim]
Rabbinic script, 1605–6
Biblioteca Palatina, Parma (MS no. 3037)

33

Scion of a distinguished Sefardic family, Moses was a gifted orator and rabbinic leader in the community of Salonika, which he represented in dealings with the Sultan. He wrote a number of important rabbinic responsa, but is best known for his series of commentaries on the Torah, the five Megillot, *Pirqei Avot*, the *Siddur*, and the Torah commentary of Abraham *Ibn Ezra. He also wrote notes on Aristotle's *Ethics* and composed a number of works in Ladino.

Almosnino's two astronomical works are *Beit Elohim* and *Shaʿar HaShamayim*, commentaries on two well-known compositions. The first is *Kadur HaOlam*, the translation of Johannes *Sacrobosco's *Sphaera Mundi*, known as *Mar'eh HaOfanim*. The second is Peurbach's *Theorica Planetarum*, *Iyyun HaKokhavim HaMeshartim*.

Moses ben Israel of Cracow (ca. 1525–72)

תורת העולה

[Torat HaOlah]
Cracow: Mordecai Katz, 1569
Yeshiva University, New York

34

Moses *Isserles was born into a very wealthy and learned family. He rapidly acquired a reputation as a master of rabbinic literature, and his interests included philosophy, Kabbalah, and astronomy. He was active in local affairs and was consulted on many matters by rabbis from around the world.

Among, Isserles' many books are *Darkhei Moshe*, a commentary on Jacob ben Asher's *Arbaʿah Turim*; the *Mapa*, an Ashkenazic supplement to Caro's *Shulhan Arukh*; a volume of rabbinic responsa; *Mehir Yayyin*, a commentary on Esther; *Torat HaHatat*, on permitted and prohibited items related to ritual matters; *Qarnei Re'em*, notes on *Mizrahi's commentary on Rashi's Torah commentary; and *Torat HaOlah*.

Torat HaOlah is a major philosophical work structured around all aspects of worship in the ancient Temple. It contains explanations for the buildings, equipment, personnel, rituals, sacrifices, forms of atonement, Temple regulations, and much more. Incorporated into *Torat HaOlah* are several passages in which Isserles mentions the astronomical works of his predecessors, for example, *Yesod Olam*, on part of which he wrote comments. But more important are the many passages in which he endows the parts of the Temple and related matters with cosmic symbolism. In Part I, chapter 5, for example, he relates the layout of the altar in the Temple to the layout of the hevens; in chapter 7, he correlates the seven gates with the seven planets. This type of astronomical association with parts of the Temple is very old, but the details in Isserles' comparisons are striking and show the importance and validity he attributed to astronomy as he knew it.

David Gans of Westphalia, Cracow, and Prague (1541–1613)

נחמד ונעים

[Nehmad VeNaim]
Jewish Theological Seminary, New York (MS no. 2563)

35

Jessnitz: Israel bar Abraham, 1743
Jewish Public Library, Montreal

36

Although David Gans is best known among Judaica scholars for his historical work, *Tzemah David*, he was a multifaceted writer. A student of both Rabbi Moses *Isserles and Rabbi Judah Lowe (the Maharal of Prague), he was also an associate of both Tycho Brahe and Johannes Kepler. Though not the intellectual equal of any of the four, Gans was uniquely positioned to absorb and integrate their teachings. He was a witness to, if not an intimate participant in, the religious and scientific developments of the sixteenth and early seventeenth centuries. According to A. Neher, his modern biographer, Gans provided the first references in Hebrew to Copernicus, Brahe, and Kepler, included the first correct map of the world in a Hebrew cosmography, and composed the first substantial Hebrew report on North America.

Containing twelve sections and over three hundred chapters, *Nehmad VeNaim* is a full cosmology. It includes discussions of various Greek accounts of the universe, a history of astronomy (Jewish and non-Jewish), and analyses of the movements of the celestial bodies and the geography of the earth. The book contains latitudinal and longitudinal tables for many locations, mathematical and geometrical discussions with illustrations, explanations of the use of the quadrant, a refutation of astrology, a discussion of the discovery of North America, and observations on the significance of many of these for Jewish religious thought and observance.

The first edition appeared about 130 years after the author's death, but the beginning of a preliminary edition was published in 1612 under the title *Magen David*. The latter was accompanied by approbations from, among others, Rabbi Yom Tov Lipmann Heller, author of *Tosafot Yom Tov* (a very famous commentary on the Mishnah), and Rabbi Solomon Ephraim Luntschitz, author of *Keli Yaqar* (an equally popular commentary on the Torah). The only known copy of the few pages of this preliminary edition is found in the Bodleian Library in Oxford.

Hayyim Vital (1542-1620)

ספר התכונה

[Sefer HaTekhunah]
Spanish rabbinic script, in various hands
Montefiore Endowment, Jews College, London (MS no. 426)   37

Jerusalem: Israel ben Abraham Bak, 1865   38
Hebrew Union College, Cincinnati

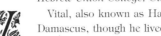

Vital, also known as Hayyim Calabrese, was born and educated in Safed. He died in Damascus, though he lived intermittently in Jerusalem, where he served as a rabbi. On the advice of Joseph *Caro, he studied with Moses Alsheikh, famed author of an extensive Bible commentary, who later ordained him. He also studied alchemy and demonstrated serious interest in dreams and visions.

Vital succeeded Isaac Luria, his main teacher in Kabbalah, when the latter died in 1572; since Luria left no written legacy, Vital's presentations of his master's teachings became the primary vehicle through which others learned them. Vital's works include a commentary on the Talmud, various letters, sermons and responsa, autobiographical pieces called *Sefer HaHezyonot*, and extensive kabbalistic writings.

Vital's ordination and service as a rabbi in Jerusalem confirm his mastery of traditional rabbinic literature, and his place in the history of Jewish mysticism is secured by virtue of his connections to Luria (though his contemporaries were not unanimous in accepting his presentations as correct). According to G. Scholem, an autograph of Vital's work on practical Kabbalah and alchemy was extant in Jerusalem as late as 1940. His *Sefer HaTekhunah*, on astronomy, extends his involvement into the scientific realm closest to Kabbalah. In it he demonstrates an awareness of Abraham *Zacuto's tables and of Abraham *Ibn Ezra's *Reishit Hokhmah*.

Vital's name is popularly associated only with Kabbalah, but his range of interests was somewhat more extensive. The integration of rabbinic learning, kabbalah, science, and what modern readers might call pseudo-science was widely accepted in the sixteenth and seventeenth centuries. His interests in astronomy, alchemy, metempsychosis (transmigration of souls), and similar matters were far from unusual for either masters of Kabbalah or humanists.

Montefiore manuscript no. 426, which contains *Sefer HaTekhunah* (reportedly copied from the author's autograph), also contains Vital's *Sefer HaLiqqutim* and several important astronomical works by other authors. Included are Zacuto's astronomical tables and a horoscope prepared by him for Damascus, February 11, 1515; a table of eclipses according to Jacob *Poel; a copy of *Al-Farghani's abridgment of the *Almagest*, transcribed from Vital's copy; a treatise on geomancy; a charm against toothache; and several other items.

Abraham ben Mordecai Azulai of Fez and Hebron (ca. 1570–1643)

בעלי ברית אברהם

[Ba'alei Berit Avraham]
Vilna: Romm, 1873   39
Yeshiva University, New York

The scientific revolution of the sixteenth and seventeenth centuries had an immediate impact on some writers, but others who were more removed from the centres of scientific research showed little or no awareness of it. Among the latter was Abraham ben Mordecai Azulai, who spent much of his life composing a series of very important Zohar commentaries that were deeply influenced by Moses Cordovera's thinking. Azulai also wrote brief commentaries on the Mishnah and the Bible, which, like his other works, were published posthumously.

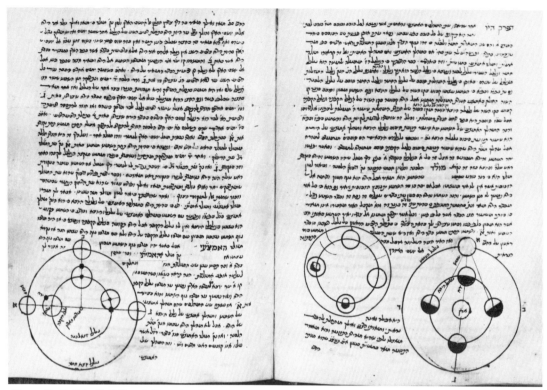

Hayyim Vital's *Sefer HaTekhunah*. Montefiore Endowment, Jews College, Collection, London, #426 (Cat. no. 37)

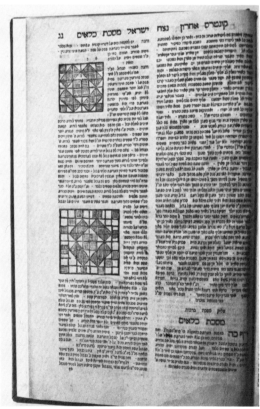

Geometric Sketches on the tractate *Kila'im*, from Israel Zamosc's *Netzah Yisra'el*, Frankfort a.M., 1741. National Library of Canada, Ottawa, Jacob M. Lowy Collection (Cat. no. 42)

The beginning of Moses Isserles' comments on the historical parts of Isaac Israeli's *Yesod Olam*, from the Cracow edition of Abraham Zacuto's *Sefer Yuhasin*. National Library of Canada, Jacob M. Lowy Collection (Cat. no. 30)

In his Bible commentary, Azulai discussed Joshua 10, the episode known popularly as "The Sun's Standing Still." He rejected the literal understanding of the story, insisting, instead, that the sun did not stop but rather continued to move on its regular path at a slower pace. Alternately emphasizing and reinterpreting specific elements in the story and simultaneously integrating them with Ptolemaic astronomical reasoning and the teachings of Abraham* Bar Hiyya's *HaMehalekhot [HaKokhavim]*, Azulai presented a cogent argument that is illustrated with two sketches of the positioning of the earth and the sun, both before and during the miracle. Though he paid lip service to the miraculous nature of the events, Azulai's approach did not rely on mystical or theological arguments. He referred to *Gersonides' statement that God acts through nature when possible, and proceeded to explain the event through the astronomical teachings known to him. His approach was scientific, if somewhat outdated.

Israel ben Moses HaLevi (Segal) of Zamosc (ca. 1700-72)

ארובת השמים
[Arubbot HaShamayim]
German cursive
Montefiore Endowment, Jews College, London (MS no. 427)

| 40 |

German cursive
Yeshiva University, New York

| 41 |

נצח ישראל
[Netzah Yisrael]
Frankfort on the Oder: Johann Koellner, 1741
National Library of Canada, Ottawa (Jacob M. Lowy Collection)

| 42 |

Born in Galicia, Israel served as a yeshiva teacher in Zamosc, an early centre of the *Haskalah* in Poland; but he exemplified the medieval Spanish synthesis of rabbinic and scientific learning and moved to Germany, where the Enlightenment was more developed. In Berlin, he was Moses Mendelssohn's teacher of mathematics and astronomy. A. Altmann, Mendelssohn's biographer, suggests that it was largely due to Israel's teaching that Mendelssohn found it possible to compose his commentary on *Maimonides' *Millot HaHiggayon*. Zamosc's involvement with the *Haskalah* placed him in conflict with the rising Hasidic movement, and his sharp criticisms of it made him even less popular in certain religious quarters.

His scientific works include *Nesah Yisrael*, which deals with the astronomical and geometric passages in the Babylonian and Palestinian Talmudim, and *Arubot HaShamayim*, never published, on astronomy. He also composed a series of commentaries on medieval philosophical classics: *Ruah Hen*, the *Kuzari, and *Hovot HaLevavot*.

Elijah ben Moses Bashyatchi (Bashyazi) of Adrianople (ca. 1420–90)
Astronomical Writings
Karaite Spanish Script, Mordecai ben Samuel, 20 Shevat 5451 (1691)
Bodleian Library, Oxford (Heb. e.12)

| 43 |

The Karaites, pious medieval Jews who rejected the authority of the rabbis and their teachings, developed an independent sect that flourished in many pre-modern centers and survives in some scattered communities. Persistent claims that they are a fundamentalistic group committed to a completely Bible-centred life style developed solely from a literal reading of Scripture reflect rabbinic bias more than historical truth. In fact, the Karaites developed their own supplementary traditions, and many of their leaders both studied philosophy and science and integrated them into their religious teachings.

One of the most important Karaite leaders was Elijah ben Moses Bashyatchi, author of *Aderet Eliyahu*, a codification of Karaite religious practice comparable in importance to Joseph Caro's *Shulhan Arukh*. Bashyatchi was a moderate Karaite who utilized many rabbinic works and even supported the lighting of Sabbath candles. This had been prohibited for many centuries because Karaite interpretation of the biblical prohibition against lighting a fire on the Sabbath included fires kindled beforehand. Bashyatchi wrote a number of astronomical works, including tables.

Later astronomic works often took the form of commentaries on Maimonides' *Hilkhot Qiddush HaHodesh* and are discussed in Chapter 6.

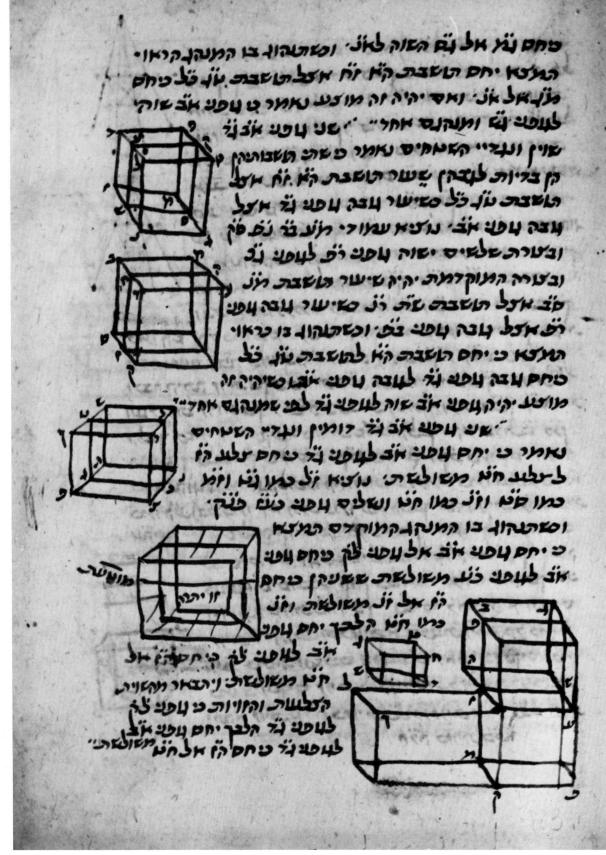

Calculating the measurements of various three-dimensional objects according to principles of Euclidian geometry, as presented in Judah ben Solomon of Toledo's *Midrash Hokhmah*. Biblioteca Palatina, Parma, MS #2769 (Cat. no. 44)

Chapter Three

Mathematics and Geometry

מִי שָׂם מְמַדֶּיהָ כִּי תֵדָע
אוֹ מִי נָטָה עָלֶיהָ קָּו:

איוב לח,ה

> ... And he [Solomon] made the "sea" [a large water basin from which priests would wash in the Temple] of metal, round, ten cubits in diameter; it was five cubits high and thirty cubits in circumference...
>
> *I* Kings 7:23

> The text says that the diameter of the "sea" was ten cubits. This is approximate, because the diameter of a circle is not exactly a third of the circumference, as the rabbis said about this in the tractate *Eruvin (14b); it [the circumference] is approximately one-seventh more than three times the diameter. Perhaps the circumference was measured inside the sea [and represents the circumference of the inner wall, not the outer wall which is greater], and this is closer to the truth.
>
> Abravanel, Commentary, *a.l.*

Passage in which Barukh ben Jacob Schick of Shklov refers to the Vilna Gaon's request to translate Euclid into Hebrew. National Library of Canada, Jacob M. Lowy Collection (Cat. no. 45)

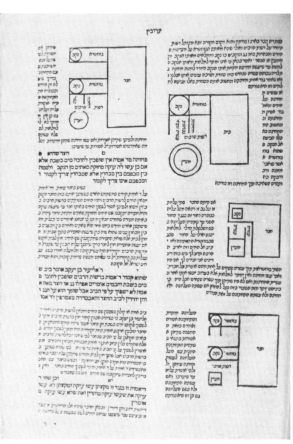

Drawings that depict various situations related to the use of *eruvin*. Maimonides' commentary to the Mishnah, *Eruvin*, Naples, 1492. National Library of Canada, Jacob M. Lowy Collection (Cat. no. 46)

Sketches depicting the relationship between a square and a circle, and the manners of measuring two-thousand cubits around cities with circular or square shapes. Babylonian Talmud, *Eruvin*, with the commentaries of Rashi and Tosafot, Pesaro, 1510 (?). National Library of Canada, Jacob M. Lowy Collection (Cat. no. 47)

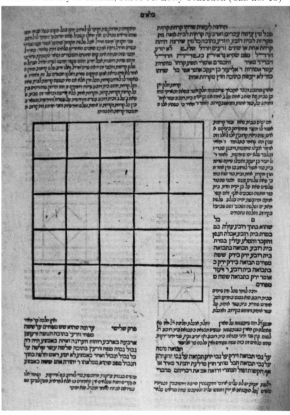

A graphic presentation of a field divided into twenty-five square sections. Maimonides' commentary to the Mishnah, *Kil'aim*, Naples, 1492. National Library of Canada, Jacob M. Lowy Collection (Cat. no. 46)

Euclid of Alexandria (ca. 300 BCE) and Judah ben Solomon of Toledo

מדרש חכמה
[Midrash Hokhmah]
Rabbinic script, fourteenth century
Biblioteca Palatina, Parma (MS no. 2769)

|44|

As is true for many ancient scientists, almost nothing is known of Euclid's life. His reputation is based on his *Elements*, a thirteen-part study of number theory and geometry, but the version in use for well over one thousand years was a later adaptation of his work. The edition produced by Theon in the fourth century became the version known throughout the world, and only in the nineteenth century did scholars discover what is believed to be the earlier recension.

*Galen, *Hippocrates, and *Ptolemy all became known through translations, which is a tribute to their universally recognized contributions to knowledge; the same is true of Euclid. In fact, the history of the translation and transmission of Euclid's work is a fascinating example of literary evolution. The *Elements* was translated into Arabic and abridged several times, and the Arabic versions were rendered into Latin by a number of famous translators. A Syriac translation that closely resembles one of the Arabic versions also exists, as do Hebrew translations preserved in several dozen complete or fragmentary manuscripts. Addenda and commentaries were also composed to the translations by *Gersonides, Afendopolo, Al-Farabi, and others.

Parma manuscript no. 2769 contains *Midrash Hokhmah*, by Judah ben Solomon of Toledo, which incorporates the author's own work with selections from Aristotle, Euclid, Ptolemy, and several other writers. The sections on geometry are accompanied by many drawings.

Barukh ben Jacob Schick of Shklov (1744–1807)

ספר אוקלידוס
[Sefer Euclidos]
Amsterdam: J. H. Munnikhuizen, 1780
National Library of Canada, Ottawa (Jacob M. Lowy Collection)

|45|

*Barukh Schick was a versatile rabbi and scientist who wrote a number of works on astronomy, medicine, and geometry. In 1777, he published Israeli's *Yesod Olam* and his own *Amudei Shamayim*, a commentary on *Maimonides' *Hilkhot Qiddush HaHodesh*, with *Tiferet Adam*, on anatomy. He also wrote *Derekh Yesharah*, on medicine and hygiene, but his most famous composition is undoubtedly the introduction to his translation of Euclid, in which he claimed that the Vilna Gaon encouraged him to translate this and similar works. The piece has been cited and embellished by two centuries of writers, who sought to justify their own predilections for scientific research through the religious teachings of the Gaon, but its reliability and the support for enlightened positions based on it have been challenged. Perhaps even more curious, but rarely, if ever, discussed, is why it was felt necessary to translate Euclid into Hebrew in the late eighteenth century, when manuscript copies of medieval Hebrew translations were available in a number of librairies.

Mishnah, Order *Zera'im*, Tractate *Kila'im*, with the commentary of Maimonides

משנה, זרעים, כלאים
Naples: Joshua Solomon Soncino and Joseph ibn Peso, 8 May 1492
National Library of Canada, Ottawa (Jacob M. Lowy Collection)

|46|

Babylonian Talmud, Order *Mo'ed*, Tractate *Eruvin*, with the commentaries of Rashi and Tosafot

תלמוד בבלי, מועד, עירובין
Pesaro: Gershom ben Moses Soncino, 1510 (?) 1515 (?)
National Library of Canada, Ottawa (Jacob M. Lowy Collection)

|47|

The Mishnah, composed at the beginning of the third century, contains the most important ancient collection of rabbinic teachings. Together with the *Tosefta* and several other sources containing ancient rabbinic material, it was the basis of all later rabbinic teachings, and was the primary subject treated in the Palestinian and Babylonian Talmudim. The latter became the supreme source of rabbinic law and lore, whose notions, interpretations, and applications have determined the course of most Jewish life for the past 1,500 years.

A number of tractates found in the Mishnah and the Talmud are devoted to matters of scientific import; two of them, *Eruvin* and *Kil'aim,* contain interesting discussions of geometry. The latter is devoted to preventing the combining of prohibited mixtures in cloth, the crossbreeding of animals, and especially intermingled crops. The former is found in the sections on religious holidays and is closely linked with the tractate on the Sabbath.

The Hebrew term *eruv* defies precise translation, because it applies to a number of unrelated objects: food left at the outer limit of the Sabbath boundary which establishes a new residence and thereby extends the boundary; commonly owned food that legally joins adjacent properties, owned by different people, into one; string or other objects attached to poles, partial fences, walls, and other items that enclose areas and make them "private domains"; and food prepared prior to a festival beginning immediately before Saturday (when cooking is prohibited) that is set aside for the Sabbath and thereby permits preparation of food for the Sabbath on the festival itself, because the preparation had begun on a weekday.

Despite the fact that three of the four types of *eruvin* are concerned with food or property lines, neither is the source of the term, though it is popularly thought to derive from the Hebrew root meaning "to mix," reflecting the legal fictions that permit the mixing of physical or temporal domains. The triliteral Hebrew root *ʿrb* is the most multifaceted one in the language (in different contexts it connotes, among other things, guarantee, sweet sounding, evening, mix, raven, and Arab). But the origin of *eruv* is likely to be found in the Akkadian *erebu,* which means "to enter," and is close to the talmudic definition of the non-edible type of *eruv* as *tzurat hapetah,* "a doorway."

Regardless of its etymology, many discussions of *eruvin* centre on matters of architecture and surveying, and are predicated on a sophisticated appreciation of mathematics and geometry. They are often abstruse – *Eruvin* has a reputation for being one of the most difficult talmudic tractates – and many commentaries clarify them with sketches. Some editions of both the Mishnah and the Talmud have preserved these sketches, which are important sources for the history of geometrical studies and the development of book illustrations.

Among the types of problems discussed are determining the precise location of the two-thousand-cubit limit outside a city (approximately one kilometre), beyond which one could not walk on the Sabbath. Circular, square, and irregularly shaped cities posed different types of difficulties, because one had to determine from what point and in which direction to measure the distance.

Kil'aim is replete with discussions of land shapes and calculating areas. The illustrations are designed to clarify the geometric or mathematical reasoning of the rabbis, and offer important insights into early rabbinic mathematics as well as into that of the commentators.

The commentary on the Mishnah was one of *Maimonides' earliest works, begun when he was twenty-three and completed in Egypt seven years later. It was composed in Arabic, translated by six or seven different medieval writers, and became one of the classics of its genre. The 1492 Naples edition of the Mishnah is the oldest one to have been preserved and contains the first edition of Maimonides' commentary. The sketches closely resemble those found in early manuscripts of his commentary.

The Pesaro edition of *Eruvin* was prepared by Israel Ashkenazi, and is heavily illustrated. Unlike many drawings in rabbinic texts, these are carefully labelled, and many of them are dissimilar to those in more recent editions.

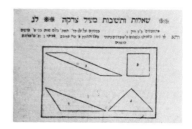

Jonah ben Elijah (Bunzlau) of Prague (1678–1718)

מאיל צדקה
[Me'il Tzedaqah]
Prague: (Sons of) Moses Bak, 1757
National Library of Canada, Ottawa (Jacob M. Lowy Collection)

48

Landsofer, a rabbi, kabbalist, scribe, and scholar, was a member of a respected rabbinic family, and was sent as an emissary to participate in anti-Sabbatian debates in Vienna. His works include an extensive volume on scribal practices (*Benei Yonah*), commentaries on several sections of the *Shulhan Arukh, and a volume of responsa entitled *Me'il Tzedaqah*.

Part of *Me'il Tzedaqah* is devoted to halakhic matters that relate to mathematics and geometry. One passage, for example, discusses the size of *gris* (a bean), important for determining matters of purity. Another deals with measuring the area of an irregularly shaped pice of cloth. In the course of answering the questions addressed to him, Landsofer criticized the accuracy of certain sketches found in older printed texts, and offered accurate measurements based on his mastery of the halakhic literature and the mathematical sciences.

Tuviah ben Meir HaLevi of Harshitz

ברורי המדות
[Berurei HaMiddot]
Prague: Sebastian Diesbach, 1807
Yeshiva University, New York

49

The analysis of the mathematical and geometric passages in early rabbinic literature was a routine part of medieval Talmud study, and was also treated in many works on mathematics and astronomy; later centuries witnessed the production of specialized works that concentrated almost exclusively on this subject. One such composition is *Berurei HaMiddot*, which deals with relevant passages in the Talmud and post-talmudic commentaries and codes, as well as principles of measurement and geometry.

Abraham ben Meir Ibn Ezra (1089–1164)

ספר המספר
[Sefer HaMispar]
Spanish rabbinic script
Montefiore Endowment, Jews College, London (MS no. 419)

50

Untitled mathematical work
Muscato ben Menahem, 1482
Bibliothèque Nationale, Paris (Heb. MS no. 1051)

51

In addition to his noteworthy efforts made in many other fields – including philology, philosophy, poetry, the Bible, astronomy, and astrology – Abraham *Ibn Ezra made important contributions to the study of mathematics. As was the case for his linguistic and astronomical work, his mathematics was drawn from Arabic sources; many historians of mathematics credit him with being the first to transfer the Indian-Arabic decimal system to Europe.

Sefer HaMispar contains a brief introduction that defines terms and is almost as much philological as mathematical. It presents the use of the Hebrew letters *Aleph* to *Tet* as the equivalents of the Arabic numerals 1 to 9 and as the components of all largers numbers (in contrast to the regular system of notation); it also introduces the use of a circle for zero. Subsequent chapters discuss a range of mathematical concepts and operations, including addition, subtraction, multiplication division, fractions, proportions, square roots, and the geometry of circles, giving examples of each.

The text of *Sefer HaMispar* in Montefiore manuscript no. 419 concludes with a problem about buying a fish that is not found in the printed edition. This manuscript, which was owned at one time by Samuel David Luzatto, also contains Ibn Ezra's *Hokhmat HaMispar*.

A related treatment appears in *Sefer HaEhad*, which combines philosophical, mystical, astrological, and mathematical discussions of the whole numbers 1 to 9. Mathematics is

Excerpts from Mordecai Comtino's mathematical and geometrical writings. Bibliothèque Nationale, Paris, MS Heb. 1031 (Cat. no. 52)

also important in Ibn Ezra's astronomical and astrological works, in his Bible commentaries, and in *Yesod Mora'*, a philosophical essay. Paris manuscript Heb. no. 1051 contains a mathematical work by Ibn Ezra accompanied by his discussions of division.

Mordecai ben Eliezer Comtino of Constantinople (fifteenth century)
Elements of mathematics and geometry
Sixteenth century
Bibliothèque Nationale, Paris (Heb. no. 1031)

52

Mordecai ben Eliezer Comtino was an intellectual leader in Constantinople. He was convinced of the importance of secular knowledge to the mastery of religious literature; a devoted follower of *Maimonides and Abraham *Ibn Ezra, he wrote commentaries on many of the latter's works. Among his students were Elijah *Mizrahi, chief rabbi of the Ottoman Empire, and Elijah *Bashyatchi and Caleb Afendopolo, leaders of the Karaites.

Comtino's scientific works include *Sefer HaHeshbon VeHaMiddot*, on mathematics and geometry; *Peirush Luhot Paras*, on the construction of astronomical equipment; *Tiqqui Keli HaTzefihah*, on sundials; *Sefer HaTekhunah* and *Ma'amar Al Liqqui HaLevanah*, on various aspects of astronomy; commentaries on Aristotle's *Metaphysics* and Ibn Ezra's *Sefer HaEhad; and *Iggeret Senapir VeKaskeset*, on kosher and non-kosher fish.

Paris manuscript Heb. no. 1031 is a collection of important and varied scientific treatises, including *Sacrobosco's *Sefer Mar'eh HaOfanim*, Ibn Haitam's *Ma'amar BeTekhunah*, *Ibn Ezra's *Reishit Hokhmah and *Keli HaNehoshet, Jacob ben Makhir's *Rovaʿ Yisrael*, and Costi ibn Luqa's *Sefer HaMaʿaseh BeKadur HaGalgal*, as well as other works on astronomy.

Elijah Mizrahi of Constantinople (ca. 1450–1526)
קיצור מלאכת המספר
[Qitzur HaMelekhet HaMispar]
Basle: 1546
Yeshiva University, New York

53

*Mizrahi, best known for his role as chief rabbi of the Ottoman Empire, was a committed student of the sciences, which he learned from Mordecai *Comtino and taught formally, in addition to Talmud and halakhah. His most famous work is his commentary on Rashi's Torah commentary, but he also composed two volumes of responsa and notes on both Euclid's geometry and *Ptolemy's *Almagest*.

Mizrahi's *Melekhet HaMispar*, published in Constantinople in 1533, was the first Hebrew mathematical work to be printed. Schreckenfuchs' Latin translation of the first section, annotated and published by Sebastian *Muenster, appeared under the title *Qitzur HaMelekhet HaMispar ... or Compendium Arithmetices, decerptum ex libri arithmeticarum institutionum magistri Eliae Orientalis*.

Muenster's edition regularly uses Arabic numbers in the Hebrew text, but Hebrew letters also serve this purpose, as in Ibn Ezra's *Sefer HaMispar. Thus, in the multiplication table, "12" is indicated by writing the letters *Aleph-Beth* from left to right; "15" by *Aleph-Heh*, and so forth. Zero is indicated by 0.

Shabbetai Sheftel Horowitz of Prague (ca. 1561–1619)
שפע טל
[Shefa Tal]
Hanau: 1612
Yehudah Elberg, Montreal

54

As is the case with many other languagues, the Hebrew letters regularly function as numbers. *Aleph*, the first letter of the alphabet, also means "one", *Yod*, the tenth letter, means "ten," and so forth. Thus, every word has a numerical equivalent, and mathematical operations can be used to suggest associations between words or combinations of words with identical or mathematically related numerical values. This notion, in tandem with the fact that biblical texts could be permuted and combined in different ways and that the names of letters could be spelled out in full (see below), gave creative interpreters almost total freedom to mold the Bible text in an infinite number of innovative ways.

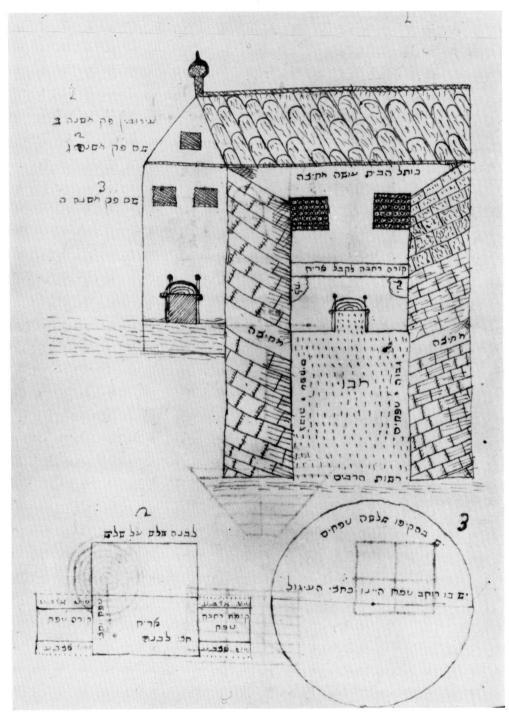

Halakhic sketches that serve as part of Isaac Noveira's graphic commentary on the first chapter of *Eruvin*. Note that the circle on the right is described as having a diameter of one and a circumference of three. Montefiore Endowment, Jews College, London, MS #53 (Cat. no. 55)

Shabbetai Sheftal Horowitz, cousin of Isaiah Horowitz (author of *Shenei Luhot HaBerit*), is best known for his daring attemps to solve certain paradoxical problems in kabbalistic thinking, but he also contributed to the numerological interpretations of the Torah by using the model of the hand and kabbalistic numerology to provide insight into various aspects of the divine nature. In biblical times, both Israelites and non-Israelites prayed by extending their hands into the air. This practice has been codified as a part of the procedure in use to this day, whereby priestly descendants of Aaron bless the people. Starting from the assumption that the priest is the vehicle through which the divine blessing is transferred and that his hands are a crucial tool in this process, Horowitz presented a detailed analysis of the numerology of the hands – human and divine – of which the following is a small sample.

A hand contains fourteen joints (two on the thumb and three on each of the four other fingers, cf. the numerological value of the two letters of the Hebrew word *yad*, "hand," which also equals 14). Two hands, 28, also represent a divine name, that is, the four letters of the tetragramaton *Yod, Heh, Vav, Heh* use twenty-eight letters when spelled out fully in Hebrew. Thus the single Hebrew letter Y= *Ywd*, which is spelled *Yod + Vav + Daleth* in Hebrew, which would be written out as *ywd + v'v + dlt* in nine letters. The other three letters of the four-letter divine name similarly contain nineteen letters, for a total of twenty-eight, and so forth.

This procedure permits discussion of God's "hands," and even allows the equation of human hands with God's power. This particular example also bears a resemblance to some forms of chairomancy. In addition to pictures of hands, *Shafa Tal* contains many drawings of kabbalistic import.

Isaac Noveira of Italy (seventeenth–eighteenth century)
Diagrams and drawings
Italian cursive

55

Montefiore Endowment, Jews College, London (MS no. 53)

Several members of the Noveira family lived in Mantua during the seventeenth and eighteenth centures, and it is possible that this Isaac is the father of Menahem ben Isaac Noveira. The latter composed several rabbinic works and prepared others for press, including *Penei Yitzhaq*, on divorce and *halitzah* procedings.

This untitled work is a collection of carefully executed sketches that illustrate a number of passages in rabbinic texts. Many pertain to specific passages in the Mishnah or the Talmud, and are so labelled. There are too few for them to be called a graphic commentary, but perhaps graphic glosses, or "tosafot," would be an appropriate designation. Among the drawings are an *eruv, a *sukkah,* and parts of the Temple, as well as astronomical drawings related to the determination of the new moon.

MATHEMATICS TEXTBOOKS

The influence of the Enlightenment encouraged greater commitment to the need for mathematical knowledge, and the composition of numerous textbooks for students. These were generally small and unattractive, and many contain collections of relevant sketches on appended fold-out sheets rather than printed at the appropriate places in the book. Sometimes they were composed in mixtures of Hebrew and Yiddish. They rarely contained innovative contributions to mathematics, as their purpose was primarily educational, but they did cover a range of subjects, including general mathematics, algebra, and geometry.

Abraham Joseph ben Simeon Wolff Menz of Frankfort on Main (eighteenth century)

ראשית למודים

[Reishit Limudim]
Berlin: 1775

56

Yeshiva University, New York

This small volume is primarily a textbook on Euclidian geometry.

Sample mathematical problems in Hebrew and Yiddish. Abraham Moses Zerah Eidlitz of Prague, *Melekhet Mahshevet*. Yeshiva University, New York (Cat. no. 57)

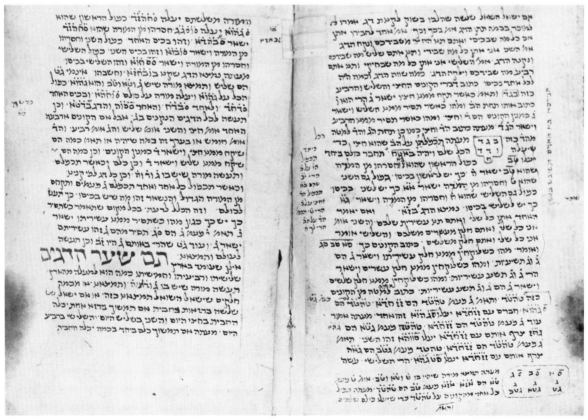

A mathematical problem about purchasing a fish that is not found in the published edition of Abraham Ibn Ezra's *Sefer HaMispar*. Montefiore Endowment, Jews College, London, MS #419 (Cat. no. 51)

Abraham Moses Zerah ben Meir Eidlitz of Prague (eighteenth century)

מלאכת מחשבת
[Melekhet Mahshevet]
Prague: 1775
Yeshiva University, New York

| 57 |

Eidlitz was raised in the home of Jonathan Eybeshuetz, and served as a member of the rabbinic court of Ezekiel Landau and, for over three decades, as the head of a yeshiva. He wrote various homilies and novellae on the Talmud; his most famous work is *Melekhet Mahshevet*, which contains Hebrew and Yiddish versions on opposite pages. The first edition appeared in 1774; the Hebrew part and an abridgment were issued separately.

David Friesenhausen (eighteenth–nineteenth century)

כליל החשבון
[Kelil HaHeshbon]
Berlin: 1796
Annenberg Research Institute, Philadelphia

| 58 |

Friesenhausen was born in Bavaria, and later lived in Berlin and Hungary. Until the age of thirty he studied classical rabbinics; he then spent a decade in scientific study. *Mosedot Tevel* expounds Copernican astronomy and contains a proof for Euclid's eleventh axiom. This textbook on mathematics and geometry also includes algebra.

Moses ben Joseph Heida of Hamburg (eighteenth century)

מעשה חרש וחושב
[Ma‛aseh Harash VeHoshev]
Frankfort on Main: Y. Koelner, 1711
Annenberg Research Institute, Philadelphia

| 59 |

Ma‛aseh Harash VeHoshev is also a mathematics textbook. The introduction is in Hebrew, but the text itself is in Yiddish and is replete with sample problems and examples.

Elijah ben Moses Gershon of Pinczow (eighteenth century)

מלאכת מחשבת
[Melekhet Mahshevet]
Berlin: I.J. Speier, 1765
Annenberg Research Institute, Philadelphia

| 60 |

Pinczow was a prolific rabbinic writer whose novellae, homilies, and legal decisions were typical of eighteenth-century Poland, but his involvement in mathematics and medicine was more unusual. *Melekhet Mahshevet*, consists of two volumes and was published several times: *Ir Heshbon*, deals with algebra; *Berurei Middot*, with geometry.

Barukh ben Jacob Schick of Shklov (1744-1807)

קנה מידה
[Keneh HaMiddah]
Shklov: Aryeh Leib ben Shayer Feibush, 1791
Yeshiva University, New York

| 61 |

In addition to his contributions to astronomy, medicine, and mathematics, *Barukh Schick also composed *Kenei HaMidah* on geometry.

TRACTATUS THEOLOGICO-POLITICUS

Continens

Dissertationes aliquot,

Quibus ostenditur Libertatem Philosophandi non tantum salva Pietate, & Reipublicæ Pace posse concedi: sed eandem nisi cum Pace Reipublicæ, ipsaque Pietate tolli non posse.

Johann: Epist: I. Cap: IV. vers: XIII.

Per hoc cognoscimus quod in Deo manemus, & Deus manet in nobis, quod de Spiritu suo dedit nobis.

HAMBURGI,
Apud Henricum Künraht. cIɔ Iɔ CLXX.

Chapter Four

General Science

הֲבָאתָ אֶל אֹצְרוֹת שָׁלֶג
וְאֹצְרוֹת בָּרָד תִּרְאֶה:

איוב לח׳, כב׳

מִי שָׁת לַמֵּעָד אַב אוֹ
מִי הוֹדִיד אַגְלֵי טָל:

איוב לח׳, כח׳

Philosophers once asked the [Jewish] sages in Rome, "If your God is opposed to idolatry, why does He not eliminate it?" They answered them, "If they worshipped an unnecesary thing, He would destroy it; but they worship the sun, the moon, the stars, and the constellations, which the world needs. Should He destroy His world because of the fools? Rather, the world functions in its normal way, and the fools who err will be held accountable [in the future].
Another comment: If someone stole a *se'ah* [e.g., a measure] of wheat and planted it in the ground, it is logical that it should not grow; but the world functions in its normal way, and the fools who err will be held accountable [in the future].
Another comment: If someone had intercourse with his fellow's wife, it is logical that she should not conceive; but the world functions in its normal way, and the fools who err will be held acountable [in the future].

Babylonian Talmud, Avodah Zara 54b
according to JTS MS no. 44830

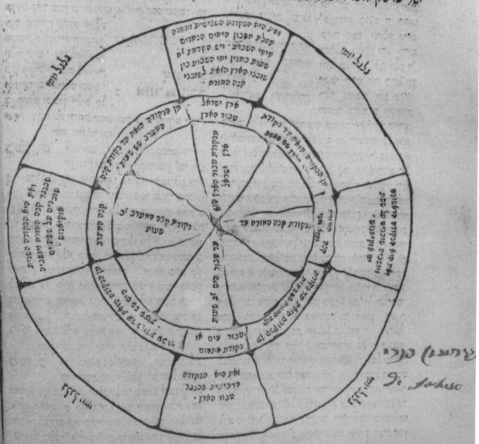

An illustrated passage from Judah Halevi's *Sefer HaKuzari*, Venice, 1594. Jewish Public Library, Montreal (Cat. no. 63)

SCIENCE IN THE PHILOSOPHICAL TRADITION

Attributed to Abraham

ספר יצירה

[Sefer Yetzira]
Rabbinic script, eighteenth century
Columbia University, New York (X893 Se3)

62

Attributed to the biblical patriarch Abraham but composed, in all likelihood, during the talmudic era (the date has been pushed back from the eighth to somewhere between the third and sixth centuries), *Sefer Yetzira* is one of the earliest and most famous works of Jewish cosmological and cosmogonical speculation. It contains esoteric wisdom about a series of issues, including the ten Sefirot, the alphabet and its mystical creative powers, and the months and signs of the zodiac. The work exists in two forms, and its extremely laconic and suggestive style, as well as the highly cryptic contents, invited further speculation. Many leading rabbinic writers tried their hand at explaining it; those who did not or whose comments were lost sometimes had others' efforts attributed to them. Some interpretations associate other cosmological and cosmogonic teachings with the book; many are repositories for extensive discourses on the subjects on which *Sefer Yetzirah* touches.

Legend had it that God used the information in *Sefer Yetzira* to create the world, and many believed that similar powers were potentially available to those who mastered the proper use of its contents. The Talmud tells how Rav Hanina and Rav Oshayah used *Sefer Yetzira* to produce a calf that would be slaughtered for the Sabbath meal every Friday (San. 65b) though it is not clear if this same book is intended. The title page of a recent reprint notes that the book provides "protection and success."

The Columbia University manuscript of *Sefer Yetzira* contains many marginal notes and a commentary attributed to Abraham ben David of Posquières. According to G. Scholem this work was composed by Joseph ben Shalom Ashkenazi, who lived around 1300. The manuscript also contains a list of phenomena arranged to match the twenty-two letters of the Hebrew alphabet, notes on *Tefillin,* a diagram of the ten *sefirot,* and a collection of *yihudim* (mystical meditations of combinations of letters) by Isaac Luria and Hayyim *Vital.

Judah ben Samuel Halevi (ca. 1075–1140)

ספר הכוזרי

[Sefer HaKuzari]
Venice: Giovanni Di Gara, 1594
Jewish Public Library, Montreal

63

Halevi, who served for part of his life as a physician, was one of the most outstanding medieval Hebrew poets; the commitment to the Holy Land that his work exhibits was rendered more poignant by his death very shortly after moving there. But undoubtedly his greatest contribution has been his *Kuzari,* written in Arabic, with a lengthy title meaning "The Book of Refutation and Proof in Defence of the Despised Faith."

On its surface, the *Kuzari* records the discussions of a Jewish sage with the king of the Khazars, who wished to choose the best religion from among Judaism, Christianity, and Islam. After examining all three, he chose Judaism. In actuality, the work is a fully developed philosophical treatise that both presents and defends Judaism. Halevi was less committed than were many of his medieval colleagues to the dominance of philosophy and science over religious thinking, and was particularly opposed to Aristotelianism. As an alternative, he stressed the experiential side of religion, downplayed analytical philosophy by arguing that God can be known only through prophecy and revelation, and limited Aristotle's contribution (which he nonetheless considered great) to the intellectual domain of his system of thought.

Judah Moscato (ca. 1530–ca. 1593), the chief rabbi of Mantua, was a typical renaissance polymath who added Greek and Jewish philosophy, astronomy, medicine, music, and Kabbalah to his classical rabbinic interests. His commentary on the *Kuzari* was one of the most important, and helped to renew interest in Halevi's work.

Abu al-Walid Muhammad ibn Rushd (Averroes) of Cordoba (ca. 1126–1198)

ספר השמים והעולם
[Sefer HaShamayim VeHaOlam]
German cursive
Montefiore Endowment, Jews College, London (MS No. 295)

64

ספר אותות השמים
[Sefer Otot HaShamayim]
Fourteenth or fifteenth century
Biblioteca Palatina, Parma (MS no. 3023)

65

Medieval philosophers and scientists knew and respected many of the Greek philosophical-scientific classics. And they wrote commentaries on them, just as they did on the *Tanakh*, the New Testament, the Mishnah, the Talmud, *Ptolemy's *Almagest*, *Galen's medical writings, and *Euclid's geometrical works. Religious documents were usually interpreted only by those who identified with their contents, but important philosophical and scientific books, and their attendant commentaries, attracted the attention of all groups. Accordingly, Ibn Rushd, one of the leading Moslem philosophers of the Middle Ages, wrote a lengthy series of commentaries on Aristotle's writings, and these, in turn, were of great interest to Jews.

Ibn Rushd, known in the west as Averroes, was *Maimonides' contemporary, and the similarities between their lives are striking. Both lived in Cordoba, Spain, Maimonides' birthplace, and both were born into highly respected families. Both were legal experts, physicians, and philosophers; and both left Spain for North Africa. There are also many differences, for Maimonides travelled eastward and settled in Egypt, while Averroes returned to Spain. Their philosophical systems differed as well but Averroes' impact on Jewish thinking was pervasive.

Averroes' approach to philosophy combined elements of both Platonic and Aristotelian thought, and often was expressed in short, medium, and long commentaries on Aristotle's books. *Sefer HaShamayim VeHaOlam* is a translation of the medium-length commentary on Aristotle's astronomical work *Peri Ouranou*, or *De coelo*, prepared by Solomon ben Ayyub HaSefaradi. Parma manuscript no. 3023 contains translations of ten scientific works, six by Averroes, including presesntations of Aristotle's *De meteoris caelestibus*, *De categoriis, De interpretatione*, and *De syllogismo*.

Moses ben Maimon of Cordoba and Fostat (1135–1204)

מורה נבוכים
[Moreh Nevukhim]
Venice: 1551
Yehudah Elberg, Montreal

66

In contrast to the biblical position that described the world as having been created by God *ex nihilo*, many ancient and medieval thinkers suggested that it was eternal. *Maimonides discussed this matter at some length, and, according to some, carefully concealed his true beliefs. On the surface, it seems that he thought, on the basis of his scientific reasoning, that the world was not eternal (Guide, II 13–24). But a tantalizing comment reveals his true opinion of the value of scientific argumentation for determining the meanings of biblical passages and ascertaining religious truth. After arguing his scientific case against the eternity of the world and for creation, Maimonides went on to add that, had he been convinced of the validity of the contrary scientific claim, he would have reconciled Genesis with it by explaining the creation story non-literally, just as he offered non-literal explanations of passages that describe God in human form (*ibid.*, 25).

Essentially, this position, which stimulated centuries of debate, assumes that the Bible does not report falsehoods and that, if it appears to report an impossibility, one should reinterpret what seems to be its expressed meaning in the light of what is known to be true. This attitude endows science and philosophy with the power to infuse the Bible with new meaning, determined largely through rational analysis.

Levi ben Gershon of Orange (1288–1344)

ספר ההקש הישר
[Sefer HaHeqesh HaYashar]
Rabbinic script, fifteenth century
Biblioteca Palatina, Parma (MS no. 2723)

In addition to his Bible commentaries and philosophical writings, *Gersonides wrote over a dozen commentaries on earlier philosophical and scientific works, including interpretations of some of Aristotle's most famous tracts. In each case, he based his analysis on reworked or abridged presentations and interpretations of Aristotle written by *Averroes, and continued the process of evaluation and debate.

Parma manuscript no. 2723 contains Gersonides' commentary on Proverbs, as well as his commentaries on three of Averroes' presentations: *Second Analytics*, a systematic discussion of the nature of scientific knowledge; *The Book of Syllogisms*, on logic; and *On the Heavens*, which deals with astronomy.

Isaac Cardoso (1604–83)
Philosophia Libera in septem libros distributa[:] In quibus omnia, quae ad Philosophum naturalem spectant, methodice colliguntur, & accurate disputantur. Opus non solum Medicis, & Philosophis, sed omnium disciplinarum studiosis utilissimum.
Venice: Bernatorum sumptibus, 1673
Columbia University, New York

Cardoso was one of the many famous Portuguese physicians and scientists who harboured a secret love of Judaism; in 1648, in the midst of an illustrious career, he abandoned his home and settled in Italy, where he wrote several multifaceted works.

Completed when he was about sixty-five years old, *Philosophia Libera* was Cardoso's final synthesis of scientific, theological, philosophical, and medical research. Its introductory history of philosophy commences with Adam and integrates the contributions of the ancient Hebrews, Chaldeans, Egyptians, Greeks, and Romans, as well as the later developers of their schools of thought.

Book *I* of *Philosophia Libera* is devoted to the principles of natural things and contains discussions of the elements, including a rejection of the theories proposed by Copernicus and the nature of the various types of bodies of water on earth – from oceans to springs – and the tides. Book *II* deals with various aspects of physics. Book *III* is devoted to astronomy, astrology, and the creation; Book *IV* deals with several topics, including other celestial bodies and phenomena: meteors, thunder, lightning, and alchemy.

Book *V* deals with plant, animal, and human life: exotic plants, the human soul, sleep, the pulse, hunger, and similar topics. Book *VI* contains 108 chapters devoted to all aspects of human existence, including physiology, psychology, prophecy, life span, mortality, resurrection, and man's varied accomplishments. Book *VII* is devoted to theology.

Among Cardoso's other works are *Utilidades del agua i de la nieve, del bever frio i caliente* (1637), on the uses of water and melted snow for medicinal purposes; *Discurso sobre el Monte Vesuvio* (1639), which describes the eruption of Mount Vesuvius in 1631; and *Las exelencias de los Hebreos* (1679), an apologia for Judaism.

Baruch (Benedict) de Spinoza of Amsterdam and The Hague (1632–1677)
Tractatus Theologico-Politicus
Hamburg: Henricum Kuenraht (Amsterdam: Jan Rieuwertsz), 1670
Jewish Public Library, Montreal

Spinoza, the son of a Portuguese Marrano who was a successful merchant in Holland, was a brilliant student of Saul Levi Morteira and Menasseh Ben Israel. His beliefs, which emerged amidst the ferment of the fertile and relatively open society of seventeenth-century Amsterdam, forced him to challenge many traditional religious ideas. Circumstances could not tolerate the freedom of expression that his radical positions required, and he was excommunicated. He wrote a number of philosophical monographs and letters, and even a scientific essay on the rainbow, but his most famous work is the *Tractatus Theologico-Politicus*.

55 / General Science

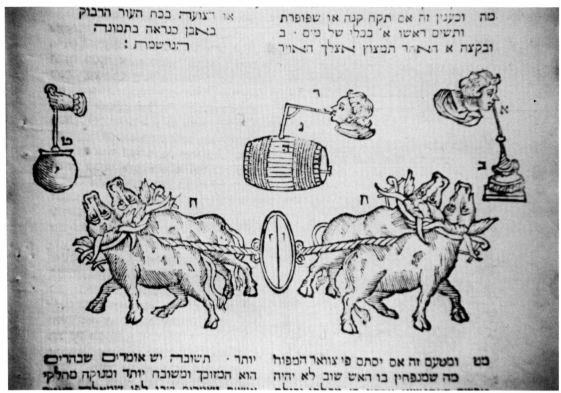

Different types of vaccuums. Tobias Cohn, *Ma'aseh Tuviah*, Venice, 1708. National Library of Canada, Jacob M. Lowy Collection (Cat. no. 73)

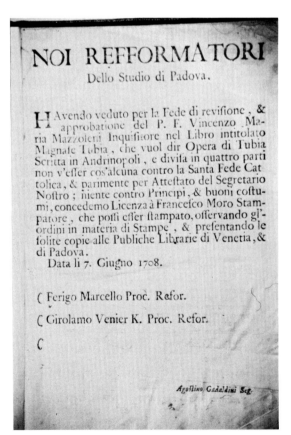

Permission to publish *Ma'aseh Tuviah*, Venice, 1708. National Library of Canada, Jacob M. Lowy Collection (Cat. no. 73)

Magnetizing a piece of iron. Tobias Cohn, *Ma'aseh Tuviah*, Venice, 1708. National Library of Canada, Jacob M. Lowy Collection (Cat. no. 73)

The *Tractatus* contains a forceful attack on Maimonides' system of interpretation and offers the alternative suggestion that the Bible be read in light of itself, not through the radical and subjective infusing of non-biblical ideas. Spinoza challenged the Mosaic authorship of the Torah and presented a developed thesis on the relationship between God and Nature, in which he claimed that God has established the world according to immutable laws and that miracles are impossible. He also advanced the belief that everything is in God, a notion very similar to one that has been popularized in many recent religious writings. Coupled with his denial of the possibility of revelation, these positions left such a gap between his teachings and those of traditional Judaism (and Christianity) that reconciliation was impossible, and Spinoza spent most of his life unaffiliated with any organized religious community.

The publication of the *Tractatus Theologico-Politicus* was so controversial that it appeared anonymously in Amsterdam in 1670 with a fictitious imprint from Hamburg.

SYNTHETIC SCIENTIFIC WORKS

Gershon ben Solomon of Arles (late thirteenth century)

שער השמים
[Sha'ar HaShamayim]
Zolkiew: Gershon Letteres, 1805
National Library of Canada, Ottawa (Jacob M. Lowy Collection)

70

Very little is known about Gershon except his thirteenth-century location in Provence and the contents of his library, both of which have been deduced from comments and sources in his book. He was once believed to be the father of *Levi ben Gershon, but this suggestion is no longer accepted.

Sha'ar HaShamayim is a textbook on the sciences, sometimes described as a scientific encyclopaedia. The first section treats of meteors, plants, animals, bees, fish, sleep, human psychology, and the like. The other two are devoted to theology and astronomy. Gershon's sources are quite widely chosen and include Hebrew translations of Plato, Aristotle, Empedocles, *Galen, *Ptolemy, *Hippocrates, *Al-Farghani, Al-Farabi, *Avicenna, *Maimonides, and *Averroes.

Abraham ben David Portaleone of Mantua (1542–1612)

שלטי הגבורים
[Shiltei HaGibborim]
Mantua: Eliezer d'Italia, 1612
National Library of Canada, Ottawa (Jacob M. Lowy Collection)

71

Abraham Portaleone was a descendant of a famous family of Italian scholars and physicians. A student of Jacob Fano and a graduate of Padua, he served as physician to Duke Guglielmo, for whom he composed two Latin books on medicine; but his most famous work is *Shiltei HaGibborim*.

On the surface, *Shiltei HaGibborim* deals with the ancient Temple and its worship services, but, in attempting to exhaust the various topics, Portaleone discussed many aspects of the related sciences. This practice, which was followed by some of the Babylonian *geonim* in their Bible commentaries and was roundly criticized by Abraham *Ibn Ezra in the introduction to his Torah commentary, often leaves the reader distracted and confused, but it served Portaleone as a useful literary device. Thus, for example, Temple music became the starting point for an analysis of music theory, the differences between priests and levites led to a discussion of the social order, and so on.

In this manner, the book treats geology, chemistry, dyeing, weights and measures, medicine, poetry, food preservation, pharmacology, architecture, philosophy, printing, and similar subjects. These are interspersed with information on military equipment and practices, the market values of gold and precious stones, and the economics of sixteenth-century Italy. *Shiltei HaGibborim* is the first Hebrew book to use European punctuation, but it is even more important as a model of the type of total integration of religious studies and the sciences that appeared in Renaissance Italy.

A *Shaduf* and an Archimedes screw in the form of a snail, both for raising water. Delmedigo, *Sefer Elim–Maʿayan Ganim,* Amsterdam, 1628–1629. National Library of Canada, Jacob M. Lowy Collection (Cat. no. 72)

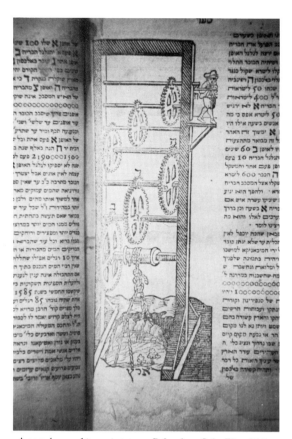

A complex machine using gears. Delmedigo, *Sefer Elim–Maʿayan Ganim,* Amsterdam, 1628–1629. National Library of Canada, Jacob M. Lowy Collection (Cat. no. 72)

The use of gears to simplify work. Delmedigo, *Sefer Elim–Maʿayan Ganim,* Amsterdam, 1628–1629. National Library of Canada, Jacob M. Lowy Collection (Cat. no. 72)

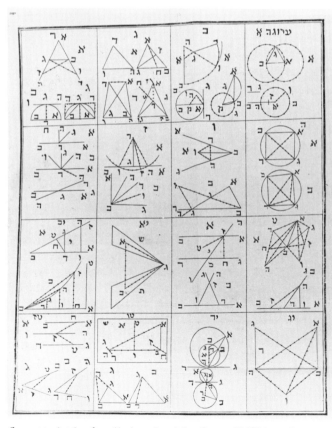

Geometric sketches from Abraham Joseph ben Simeon Wolff Mentz, *Reishit Limudim,* Berlin, 1775. Yeshiva University, New York (Cat. no. 56)

Joseph Solomon ben Elijah Delmedigo of Candia (1591–1655)

ספר אלים-מעין גנים
[Sefer Elim - Ma'ayan Ganim]
Amsterdam: Menasseh Ben Israel, 1628–1629
National Library of Canada, Ottawa (Jacob M. Lowy Collection)

72

Known best by the acronym Yashar *(Yoseph Shelomo Rofe')* MiCandia, Delmedigo was the learned son of the rabbi of Crete and the scion of a distinguished rabbinic family. He studied at the University of Padua, but spent much of his life moving from one intellectual centre to another and reaping knowledge from the masters he met in each. His associates included Ari ibn Rahmadan, an Egyptian mathematician; Leone Modena, the controversial Italian rabbi; Prince Radziwell of Vilna, whom he served as private physician; Galileo, whose astronomical thinking influenced his own; and Menasseh Ben Israel, the first Hebrew printer in Amsterdam and the publisher of *Sefer Elim*. Delmedigo lived as well in Constantinople, Rumania, Prague, Hamburg, and Frankfort am Main.

A true polymath, Delmedigo was comfortable in many languages and wrote on most of the sciences, including mathematics, geometry, medicine, alchemy, astronomy, geography, and chemistry. Unfortunately, many of his compositions have been lost. He also devoted attention to classical religious issues, including Kabbalah.

Delmedigo supported the astronomical innovations of *Copernicus, Brahe, and Galileo against the long-popular teachings of *Ptolemy; he praised Kepler as the greatest mathematician of the age. He discussed the reasons for the contemporary changes in astronomical thinking, but still devoted much attention to Ptolemy's *Almagest*, the classic astronomical work that had engaged and guided scholars for well over a dozen centuries.

Delmedigo applied his scientific knowledge to biblical and post-biblical writings alike. Like many other authors, he too noted the inconsistency between the report that King Solomon's Temple basin (*yam*) had a circumference three times its diameter and the geometric reality (*I* Kings 7:23). As well, he approached the descriptions of Noah's rainbow (Gen. 9:13), Elisha's floating an axhead (*II* Kings 6:5–7), and Ahaz's sundial (Isaiah 38:7–8) with scientific sophistication and skepticism.

Sefer Elim, the general title for Delmedigo's published scientific writings, properly refers to the first of five treatises, though only *Ma'yan Ganim* carries an additional title page. *Sefer Elim* proper is a series of responses to questions about scientific and related philosophical concerns. *Ma'ayan Ganim* contains *Sod HaYesod*, on mathematics; *Huqqot HaShamayim* and *Gevurot Adonai*, dealing with geometry, geography, and astronomy; and *Ma'ayan Hatum*, a collection of responses to various questions, many of which pertain to scientific issues and to occasional contradictions between science and the Bible. Approbations of rabbis Leone Modena, Simhah Luzzatto, Nehemiah Saraval, and Jacob Levi, which were based on a reading of *Sefer Elim*, appeared at the beginning of *Ma'ayan Ganim*. In the 1864 edition, they were transferred to the beginning of the volume and applied to all of it. *Sefer Elim – Ma'ayan Ganim* is the first printed Hebrew book accompanied by extensive scientific illustrations.

Tobias ben Moses Cohn of Metz, Cracow, Frankfort on the Oder, Padua, Turkey, and Jerusalem (1658–1729)

מעשה טוביה
[Ma'aseh Tuviah]
Venice: Nella Stamperia Bragadina, 1708
National Library of Canada, Ottawa (Jacob M. Lowy Collection)

73

Jesnitz: I.B. Abraham, 1721
Jewish Public Library, Montreal

74

Tobias (Tuviah) Cohn studied Talmud and other classical religious subjects in Poland, and later moved to Italy to complete at Padua (under the direction of Solomon Conegliano) the medical training he had begun at Frankfort on the Oder. He returned to Poland briefly, but soon left for Turkey, where he became an accomplished physician in the courts of five successive sultans and completed *Ma'aseh Tuviah*. Several years before his death, Cohn emigrated to Jerusalem to devote himself to religious studies.

Divided into four volumes (in five parts), *Ma'aseh Tuviah* is, in effect, a textbook of the sciences, especially medicine. But, following in the medieval and renaissance traditions, it discusses many religious themes as well as the elements, astronomy, botany,

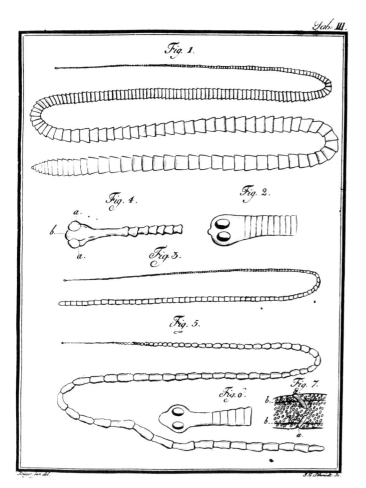 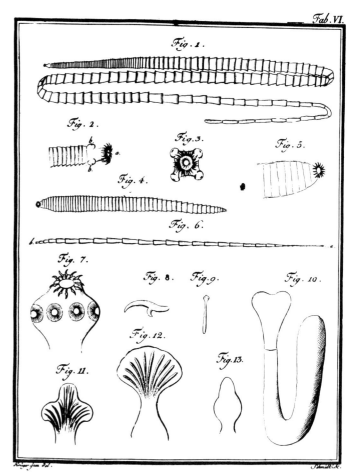

Sketches of worms. M.E. Bloch, *Abhandlung von der Erzeugung der Eingeweidewuermer*. The Wellcome Institute Library, London (Cat. no. 74)

cosmography, human biology, meteorology, magnets, medicine, pharmacology, and obstetrics.

Cohn was quite open about the need to experiment and to explore for new scientific information rather than to rely on rabbinic tradition, and he was one of the first doctors to accept Harvey's theory of the circulation of the blood. He was, moreover, strongly opposed to magical beliefs, including superstitions and the use of amulets. But one should not assume that he was merely a modernizing reformer, as he was equally negative about *Copernicus and his astronomical revolution, calling him "the first-born of Satan." This objection was based on a thorough knowledge of the new astronomy, which is explained and illustrated in Ma'aseh Tuviah, but drew its support from traditional understanding of certain biblical passages that, in his estimation, invalidated Copernicus's reasoning.

Ma'aseh Tuviah is the best-illustrated Hebrew medical work of the pre-modern era, and its most famous picture is undoubtedly the comparison of a human body and its internal organs with different parts of a house. The roof is paired with the scalp, the windows with the eyes, the bladder with the plumbing system, and so on.

SPECIALIZED SCIENTIFIC STUDIES

Marcus (Mordecai) Elieser Bloch of Berlin (1723–1799)
Abhandlung von der Erzeugung der Eingeweidewuermer und den Mitteln wider dieselben...
Berlin: Siegismund Friedrich Hesse, 1782
Wellcome Institute Library, London

75

Naturgeschichte der Auslaendischen Fische
Berlin: 1785–95
The Newberry Library, Chicago

76

*Bloch was Moses Mendelssohn's friend and physician. He wrote a number of medical works and a treatise on worms, but his books on fish, accompanied by beautiful colour plates, are the most extensive and the best known.

Barukh ben Judah Loeb Lindau of Hanover (1759–1849)

ראשית לומדים
[Reishit Limudim]
Berlin: Hevrat Hanokh Nearim, 1789
Jews College, London

77

Barukh was a mathematician and author of texts on general science. His *Reishit Limudim* was issued in two parts. The first, dealing with physics and geography, appeared in 1789. It contains sections on astronomy, the earth's movements and climatic zones, the atmosphere, animals, plants, rocks, man, and new geographic discoveries. The second volume, issued posthumously, is primarily concerned with natural philosophy.

Emanuel Mendes Da Costa of England (1717–1791)
A Natural History of Fossils
London: 1757
Jews College, London

78

Da Costa, from a well-known Marrano family from Portugal that established itself in England, was an outstanding scientist. His life was marred by various legal and financial setbacks, but he remained an active member of numerous learned societies, corresponded with many scholars, and wrote a number of books on scientific themes. Among the most prominent are *Elements of Conchology, A Natural History of Fossils,* and *Historia Naturalis Testaceorum Britanniae,* all devoted to the study of shells.

METEOROLOGY

Joseph (?)

ספר היסודות

[Sefer HaYesodot]
Spanish rabbinic script, fifteenth-sixteenth century (?)
Montefiore Endowment, Jews College, London (MS no. 433)

79

Anonymous

ספר רעשים ורעמים

[Sefer Ra'ashim VeRa'amim]
Italian scripts, seventeenth-eighteenth century
Wellcome Institute Library, London (Heb. no. 17)

80

Over the centuries, many Jewish writers commented on the weather and the proper reading of signs signaling impending changes in it. Sailors and merchants, who risked death or financial ruin with virtually every embarkation, were particularly sensitive to the capricious nature of the weather, and scholars were anxious to explain what was central to almost all aspects of daily existence. Some attempted formal works of meterology.

Sefer HeYesodot, in nine chapters, deals with snow, clouds, cold, hail, wind, lightning, the rainbow, and similar phenomena. It is bound with Judah al-Harizi's translation of *Iggeret HaMusar*, attributed to Aristotle. The very brief *Sefer Ra'ashim VeRa'amim*, accompanies a series of other brief, folksy works of a religious or quasi-scientific nature: Hebrew texts of remedies, special prayers for the month of Elul, and astrological notes, as well as Italian information about the 1348 pestilence in England.

ARCHAEOLOGY

Pre-modern rabbinic sources record a number of casual discoveries of buildings, texts, and artifacts from ancient times, and it is interesting to note how, and the ease with which, they were absorbed. Many ancient traditions refer to the burying of sacred texts and utensils in the hope of their being exhumed at some future time; occasional notes on accidental discoveries confirm an awareness of other potential recoveries. Rabbinic writers discussed, for example, the remains of old buildings on the Temple Mount, the locations of various graves and holy sites, and the precise identities of biblical sites. Ancient coins were among the most popular pre-modern archaeological topics. These contained the paleo-Hebrew script used by the Samaritans, which was discussed in the Talmud and believed by some people to be the original script of the Torah.

Moses ben Nahman of Gerona (1194–1270)

ביאור על התורה

[Biur al HaTorah]
Venice: 1545
National Library of Canada, Ottawa (Jacob M. Lowy Collection)

81

Nahmanides, a leader of Spanish Jewry who fled to Israel after successfully defending Judaism in public debates in 1263, wrote one of the most outstanding medieval commentaries on the Torah. In it, he demonstrated his mastery of rabbinic literature, his commitment to using incipient Kabbalism in Torah interpretation, and his ability to offer a radically different and seemingly more traditional approach to Torah interpretation than that of *Maimonides. Nahmanides respected his Egyptian predecessor and his work, but he opposed the degree of attention and importance that he and his followers gave to philosophy and science, and regularly offered his own alternatives.

At the very end of his commentary on Deuteronomy, Nahmanides added a brief note in which he reported on an ancient coin he saw in Acco. On one side he perceived a flowering almond staff, and on the other a glass container. On both sides were words in old Hebrew script that could be read by Samaritans: *sheqel hasheqalim* and *yerushalayim haqedoshah*. He also discussed the weight of the coin and its implications for various halakhic discussions in the Talmud.

Azariah ben Moses De Rossi of Mantua (ca. 1511–ca. 1578)

מאור עיניים

[Meʿor Einayyim]
Mantua: M. S. Ghirondi, 1573–5
National Library of Canada, Ottawa (Jacob M. Lowy Collection)

82

Azariah de Rossi, a sixteenth-century Italian who is often credited with being the father of *Wissenschaft des Judentums*, is best known as the author of *Me'or Einayim*. This highly controversial book which was condemned by Rabbi Judah Lowe (the Maharal of Prague) but praised by Rabbi Zvi Hirch Chajes, introduced or advanced many critical attitudes toward traditional Jewish texts. It drew attention to Philo and the Apocrypha as serious concerns of rabbinic learning, discussed at length the history of non-fundamentalistic reading of rabbinic midrash, and criticized the presentations of many earlier and respected rabbinic historians.

Among its many chapters is a section in which de Rossi discussed the coin described by *Nahmanides. On the basis of other similar coins he saw in Ferrara, he suggested that the script on Nahmanides' coin must have read *sheqel yisrael*. These and similar coins were also discussed even earlier by Rabbi Moses ben Isaac ibn Al-Ashqar (1466–1542) in his responsa, *She'eilot uTeshuvot MaHaRaM Al-Ashqar*, no. 64.

A shekel from 68 CE like that discussed by Nahmanides and De Rossi. Redpath Museum, McGill University, Montreal, R.B.Y. Scott Collection (Cat. no. 81 and 82)

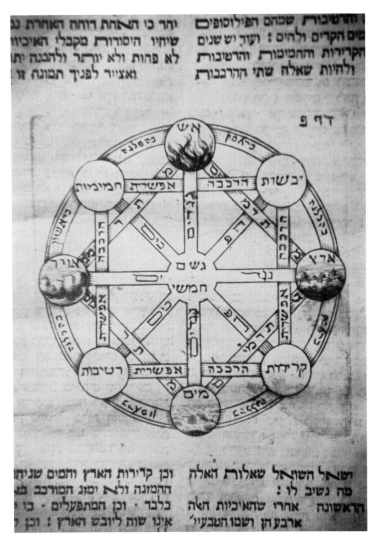

Graphic representation of the four elements and the four humours with intersecting links. Tobias Cohn, *Maʿaseh Tuviah*, Venice, 1708. National Library of Canada, Jacob M. Lowy Collection (Cat. no. 73)

Tobias Cohn, physician and scientist, author of *Ma'aseh Tuviah*, Venice, 1708. National Library of Canada, Jacob M. Lowy Collection (Cat. no. 73)

Chapter Five

medicine

וַיֹּאמֶר אִם שָׁמוֹעַ תִּשְׁמַע לְקוֹל
ה' אֱלֹהֶיךָ וְהַיָּשָׁר בְּעֵינָיו תַּעֲשֶׂה
וְהַאֲזַנְתָּ לְמִצְוֹתָיו וְשָׁמַרְתָּ כָּל חֻקָּיו
כָּל הַמַּחֲלָה אֲשֶׁר שַׂמְתִּי בְמִצְרַיִם
לֹא אָשִׂים עָלֶיךָ כִּי אֲנִי ה' רֹפְאֶךָ:

שמות טו, כו

The best of doctors [should go] to hell.

Mishnah, Qiddushin, end

He does not fear the disease, he eats the food of the healthy, and he does not break his heart to God. And sometimes he kills people and has in his power to heal the poor and he does not.

Rashi, commentary, a.l.

The Torah has permitted physicians to treat the sick; indeed, it is part of the [biblical] requirement to save a life. One [with this opportunity] who refrains from fulfilling this commandment is a murderer ...

One should not try to treat someone unless he is competent; and in the presence of a greater expert, he should yield to the latter ...

One who treated a patient without the permission of the court [and injured him] is required to pay [damages], even if he is competent; but if he treated a patient with the permission of the court and erred in the treatment, thereby injuring him, he is free of any charges in a human court but is liable in the heavenly court ...

A physician may not charge for his wisdom or study, but he may charge for the trouble and time lost [from other activites] ...

One who possesses drugs needed by a patient may not raise the price above what they are actually worth ...

Joseph Caro, Shulhan Arukh,
Yoreh De'ah, 336

Cum priuilegio Pontificis maximi Leonis decimi: τ Francisci christianissimi Francorum regis.

Omnia opera ysaac in hoc volumine contenta: cum quibusdam alijs opusculis·

Liber de definitionibus.
Liber de elementis.
Liber dietaru vniuersaliuz: cum cōmēto petri hispani.
Liber dietarum particularium: cum cōmento eiusdem.
Liber de vrinis cum commento eiusdem.
Liber de febribus.
Pantechni decem libri theorices: et decem practices:

cum tractatu de gradibus medicinarum constantini.
Viaticum ysaac quod constantinus sibi attribuit.
Liber de oculis constantini.
Liber de stomacho constantini.
Liber virtutum de simplici medicina constantini.
Compendium megatechni Galeni a constantino compositum.

Cum tabula τ repertorio omnium operum et questionum in cōmentis contentarum.

Isaacus Judaeus, *Omnia Opera*. Jewish Public Library, Montreal (Cat. no. 87)

Hippocrates of Cos (ca. 460–370 BCE) and Galen of Pergamum and Rome (ca. 129–ca. 200)

ספר אפורימוש מאבוקראט

[Sefer Aforismos MeAbucrat]
Rabbinic script, fourteenth century (?)
The Vatican (Ebr. no. 368)

83

Hippocrates, a famous Greek physician of late biblical times, is the reputed author of a collection of works that, in all likelihood, contains much that originated with others. Accordingly, it is more appropriate to speak of the *Corpus Hippocraticum* than of Hippocrates' personal writings. This collection of over one hundred compositions covers a wide variety of medical fields; diagnosis, prognosis, anatomy, physiology, surgery, therapy, gynaecology, mental illness, and medical ethics. Perhaps most famous is the Hippocratic oath, still administered to many physicians upon completion of their studies, though in some places it has been replaced by one of the versions of a similar text associated with *Maimonides.

The aphoristic literature contains collections of brief statements about medical practice. Though associated with Hippocrates, it evolved continually from ancient to modern times. Virtually all pre-modern physicians studied it in one form or another, and many tried their hand at interpreting or improving upon it. Many selections were incorporated into general medical works or collections of individual authors' compositions, or were the subject of formal commentaries by, for example, *Delmedigo, *Avicenna, and *Maimonides.

Galen may be the most famous medical figure of all time. He was born in Asia Minor, and trained there and in Alexandria; he practised in several centres, including Rome, where he treated many of the Roman gladiators and learned much from his work in keeping them in shape and caring for their severe wounds. He is alleged to be the inventor of Theriac, a mixture of many ingredients (eventually over one hundred) that began as a cure for snakebite and, after centuries of adaptation, came to be used as both a prevention and a cure for all illnesses. He is also credited with a commentary on Hippocrates's *Aphorisms*.

Vatican manuscript Ebr. no. 368 contains fifteen compositions, and is a good example of the type of anthology in which many of the medieval medical and scientific texts are preserved. Included are three texts on the preparation of salves and ointments; one each on surgery, agriculture, the pulse, fevers, ritual slaughter, astronomy, and astrological cures; several on urology and diagnostics; and Galen's commentary on Hippocrates' *Aphorisms*.

Babylonian Talmud. Order *Nashim*, Tractate, *Yevamot*

תלמוד בבלי, נשים, יבמות

Venice: Daniel Bomberg, 1522
National Library of Canada, Ottawa (Jacob M. Lowy Collection)

84

One of the *Talmud's many concerns is circumcision, including how and when this essential rite was to be performed and what conditions might necessitate its postponement or cancellation. Fever or jaundice, for example, was an accepted reason for delaying the operation. Fear that the child could not recover from the surgery was grounds for not performing it at all because, despite its importance, fulfillment of the commandment was not deemed worth risking or sacrificing the baby's life.

Included among the children thought to be at risk from circumcision were those whose brothers or cousins had died from it. Specific concerns centred on children of the same mother or of sisters who were known to be at risk because they might have a specific blood disease. This discussion, which recognizes both the sex-linked quality of the disease and the fact that it is prominent in certain families, appears to be the earliest discussion of hemophilia.

Bomberg's was the first complete edition of the Babylonian Talmud and became the page model for virtually all subsequent editions. This volume is part of a complete set of the Venice edition in the Jacob M. Lowy Collection, one of the few complete copies in the western hemisphere.

Isaac Judaeus of Egypt and Kairouan (ninth–tenth centuries)
Diete universales and Diete particulares
Italy: second half of thirteenth century
McGill University, Montreal (Osler Library no. 7626)

85

De particularibus diaetis libellus
Padua: Matthaeus Cerdonis, 23 March 1487
Hebrew Union College, Cincinnati

86

Omnia opera ysaac in hoc volumine contenta quibusdam aliis opusculis. De definitionibus, de elementis, liber diaetarum univ. cum commento Petri Hispani, de urinis, de febribus, Pantechni decem liri theorices et decem pract. cum tractatu de gradibus medicinarum Constantini, viaticum Ysaac, de oculis, de stomacho, liber virtutum de simplici medicina, compendium megatechni Galeni a Constantino compositum
Lyons: 1515
Jewish Public Library, Montreal

87

Known in Arabic as Abu Ya'qub Ishak ibn Sulaiman al-Israeli, Isaac is reported to have lived for over a century, and spent most of it in Kairouan (North Africa) in the service of the Caliph. He is believed to have been the student of Ishak ibn Amram of Baghdad and the teacher of, among others, Dunash ibn Tamim.

Israeli's philosophical writings qualify him as one of the earliest Jewish Neoplatonists, and his influence on some medieval philosophers was considerable. *Maimonides, however, had little use for his philosophical work and, in a much-quoted letter to Samuel ibn Tibbon, referred to him as "only a physician."

Israeli's medical works were composed in Arabic, from which Hebrew and Latin translations were made. Among them are works on foods and simple remedies, urine, and fevers; the full title of his collected works, *Omnia Opera*, is essentially a table of contents. Some of the many compositions attributed to him were, in all likelihood, written by others.

Many of Isaac's philosophical works were translated into Latin by Gerard of Cremona. Constantinus Africanus, an eleventh-century convert to Christianity, who helped make Arabic medicine known in Europe, translated his medical writings; their popularity is evidenced from the fact that they were considered standard texts in several European medical schools for centuries.

Avicenna [Abu 'Ali al-Hussein ibn Abdallah Ibn Sina of Bagdad] (ca. 980–1037)

אל קאנון-הספר של אבן סינא
[Al-Kanon, HaSefer Shel Ibn Sina]
Spain or Italy: Spanish rabbinic script, late fourteenth century
Biblioteca Universitaria, Bologna (MS no. 2297)

88

[HaKanon HaGadol]
Joab ben Abraham ben Joab of Tivoli, 1334
The Vatican (Ebr. no. 564)

89

[HaKanon HaGadol]
Naples: Azriel ben Joseph Ashkenazi Gunzenhauser, 9 November 1491 (1492?)
National Library of Canada, Ottawa (Jacob M. Lowy Collection)

90

*Avicenna was an important contributor to medieval thought, but his impact on the history of medicine has been even greater. *Al-Qanun fi al-Tibb*, his monumental encyclopaedia of medical knowledge, is divided into five books, and includes lengthy discussions of the human body, principles of medicine, drugs, diseases of individual organs and of the whole body, and pharmacology. It incorporates many of the teachings of Avicenna's predecessors, including *Galen and *Hippocrates. Similar in many ways to the work of Rhazes, it was the most thorough and extensively used medical work in the Middle Ages, and is still studied and followed in many Moslem countries. It was rendered into Latin in the twelfth century by Gerard of Cremona, who also translated the *Almagest*.

The Hebrew text of the *Canon* has been preserved in many manuscripts. Bologna manuscript no. 2297 contains Rabbi Zerahiah ben Isaac ben Shaltiel's Hebrew translation of the first book of the *Canon*, which deals with general medical matters. The illustration, which has been dated to the early fifteenth century, is Italian; while it purports to depict Avicenna, it is actually a picture of a contemporary physician.

Manuscript Ebr. no. 564 contains book one of the *Canon*, translated by Zerahiah ben Isaac ben Shaltiel also. Bound with it is a copy of Bernardo di Gordon's *Shoshan HaRefu'ah*, translated by Yekutiel ben Solomon.

The Naples edition, the only Hebrew version of the *Canon* ever published, contains the translations of Joseph ben Joshua Lorki (Book *I*) and Nathan HaMeʿati (Books *II–V*). It has the distinction of being the first scientific work published in Hebrew.

Anonymous

שו"ת על קצת מהאפן הא' מהספר הא' לבן סינא
She'eilot uTeshuvot ʿal Qetzat HaOfan HaRishon MeHaSefer HaRishon LeBen Sina]
Spanish cursive, sixteenth–seventeenth century
Wellcome Institute Library, London (Hebrew MS no. 12)

91

Important legal, philosophical and medical works were often accompanied by commentaries and augmented by responsa (*she'eilot uteshuvot*) in the Middle Ages. As one of the seminal medical texts of the era, *Avicenna's *Canon* was treated similarly.

This manuscript, composed of five whole or partial works, among them extracts from the *Canon*, contains a number of medical responsa. Several on fevers cite *Galen and Avicenna; others are extracted from Pedro Hispanus' commentary on Galen. Another section, beginning on page 21r with the introductory phrase adopted as its title, contains questions and answers on Book *I* of Avicenna's *Canon*.

Abdu-l Mutarrif Abd al-Rahman Ibn Waafid of Toledo (died 1075)

ספר מראשות הראש
[Sefer Mera'ashot HaRosh]
Oriental rabbinic cursive, Abraham ben Reuben Amilabi, 1391
Biblioteca Palatina, Parma (MS no. 2116)

92

Sefer Mera'ashot HaRosh, known in Latin as *Liber de cervicalibus capitis*, was translated by Judah ben Solomon Nathan, and was provided with appendices by the translator and with indices by the scribe. It deals with a range of medical matters.

Moses ben Maimon of Cordoba and Fostat (1135–1204)
Aphorismi secundum doctrinam Galeni
Bologna: Benedictus Hectoris, 29 May 1489
Hebrew Union College, Cincinnati

93

*Maimonides, a philosopher, talmudist, and legal authority, also served as personal physician to Saladin. He was educated in Cordoba until the age of thirteen (when his family was exiled) and perhaps later in Fez. The source of his medical training is uncertain, but he appears to have been a functioning physician by the time he arrived in Fostat, Egypt, in 1165.

A respected religious leader, Maimonides refused to be paid for his services, preferring to support himself and his family from his medical work; he was, by all accounts, a dedicated and extremely competent physician. His very demanding routine – which included one small midday meal; daily ministering to the Sultan's family, harem, and officials; and prescribing for the sick and needy – has become one of the most frequently cited aspects of his biography. These pressures, even if not present throughout his entire adult life, make even more impressive his profound and prolific literary output, which included a commentary on the Mishnah; the most all-inclusive halakhic code ever written, the *Mishneh Torah;* numerous rabbinic responsa; the most seminal Jewish philosophical work of the Middle Ages, *Guide for the Perplexed;* and a series of important medical books.

Composed in Arabic, but translated into Hebrew, Latin, and other languages, *Aphorisms According to Galen* contains about 1,500 pieces of medical information in twenty-

A page from the Arabic text of *Regimen Sanitatis*. Wellcome Institute Library, London, Arabic WMS Or. 27 (Cat. no. 94)

The beginning of Maimonides' *Regimen Sanitatis*, 1518. Hebrew Union College, Cincinnati (Cat. no. 95)

Moses ben Maimon, *Regimen Sanitatis*, 1518, with marginal notations believed to have been written by Martin Luther. Hebrew Union College, Cincinnati (Cat. no. 95)

five chapters devoted to, among other concerns, the organs of the body, the humours, the pulse, the causes and treatments of diseases, fevers, the different means of purgation, the diseases of women, surgery, exercise, and food and drink. Far from a slavish repetition of the Galenic literature known to him, Maimonides' work contains criticisms, corrections, and additions, as well as attempts to differentiate between authentic and spurious comments attributed to the ancient master.

Abu° Imrar Musa b. °Ubaid-Allah ibn Maymun al-Isra'ili al-Qurtubi (Maimonides)
Tadbir-s-sihha: Tadbir Yu°tamad °Alayh fi Shifa' Amrad Hadathat li Mawlana, al-Maqala fi Tadbir-s-sihha-l-afdaliya [Regimen Sanitatis]
Eighteenth century
Wellcome Institute Library, London (WMS Or. 27)

94

Tractatus Rabbi moysi de regimine sanitatis ad Soldanum Regem
Augustae Vindelicorum: Grim and Wirsung, 9 July 1518
Hebrew Union College, Cincinnati

95

Tadbir-s-sihha, or *Regimen Sanitatis*, as it is popularly known, was written between 1198 and 1200 for Saladin's son, who suffered from mental and physical problems. The first chapter deals with personal medical matters and diet; the second with medical self-help practices to be used in the absence of a physician. The fourth contains a great deal of information on general matters of health and the daily behaviours that influence it, such as bathing, sexual activity, heat, and cold. Chapter 3 is devoted to the Sultan's personal situation, and provides information on Maimonides' understanding of the relationship between physical and mental health and his treatment of a famous patient.

This copy of the Latin translation of *Tadbir-s-sihha* contains marginal notes believed to have been written by Martin Luther.

Shem-Tov ben Joseph ibn Falaquera [Palqera] of Spain (ca. 1225–1295)

צרי היגון
[Tzeri HaYagon]
Cremona: Vincenzo Conti, 1557
National Library of Canada, Ottawa (Jacob M. Lowy Collection)

96

Falaquera, a philosopher and scientist, was the author of many varied but less than completely original compositions. A strong supporter of *Maimonides, he composed a commentary on parts of the *Guide for the Perplexed* and a defence of its author from Jewish detractors. With Maimonides, Falaquera believed in spiritual salvation through the intellect and in the necessity of allegorical interpretation of various biblical passages that on the surface did not accord with scientific teachings.

Falaquera's *Iggeret HaViquah* is a dialogue between a religious scholar, who defends philosophical and scientific considerations of Jewish beliefs, and a pietist opposed to such synthesis. This apology for philosophy is designed for the layman, not for philosophers, and is entertaining as well as informative.

Among Falaquera's medical or quasi-medical writings is *Tzeri HaYagon* a composition described alternatively as psychology or ethics. It deals with the difficulties of coping with grief, and supposedly was modelled on Galen's "On Dispelling Worry." The first edition was published in 1557, and its popularity is evident from the large number of subsequent editions.

Anonymous

ספר תולדות
[Sefer Toladot]
Spanish rabbinic script, fifteenth century
Montefiore Endowment, Jews College, London (MS no. 440)

97

Midwifery was a respected practice in medieval times. The frequent complications and problems associated with child birth required both theoretical and practical knowledge. *Sefer Toladot*, a dialogue between an unknown "Dinah" and her father, purports to provide

שלחן ערוך

מטור יורה דעה הנקרא בית יוסף

חברו הגאון מופת הדור החכם השלם מהר"ר יוסף קארו נר"ו בן מהר"ר אפרים קארו זצ"ל אשר אור תורתו
זרחת כאור היום בעיר צפת תוב"ב ומעיני תורתו נפוצות ביהודה ובישראל נודע שמו :   וחבר
הספר הזה קיצור מחבורו הגדול אשר עשה ־ על הארבעה ־ טורים אשר קראם
   אשר בסכה מעשיו הגיד וכל יקר ראתה עינו כדי שכל
מבקש ה' ימצא מבוקשו בנקל כל דין ודין על
מתכונתו באין אומר ואין דברים והכין לכל מטה־ושלחן וכסא
ומנורה אשר לאורו ילכו בטח :   כי כן משנת
רבי יוסף קב ונקי :

נדפס בבית מסי זואן גריפו בחדש אדר חמשת אלפים שכ"ז ליצירה
פה ויניציאה הבירה

The title page of Joseph Caro's *Shulhan Arukh*, Venice, 1567. Jewish Public Library, Montreal (Cat. no. 100)

this information, but it also appears to be a guide for a young wife who is concerned about the bearing of children.

The introduction indicates that Dinah approached her father out of pious and biological concerns and asked him to provide answers to her medical and philosophical questions, as well as practical information. He agreed, and a lengthy dialogue followed. The conclusion notes that Dinah then left her father and cohabited with her husband, Job, producing children who populated the entire earth.

The borrowing of names from early biblical times (including the undated Job, who is made a contemporary of the patriarchs by many rabbinic interpretations) gives the work an aura of antiquity it would not otherwise have. In reality, it is a Hebrew adaptation of an early Latin composition that has a long and complicated history of transmission and confused identification.

Bruno da Lungoburgo (thirteenth century)

ספר הכריתות
[Sefer HaKeritot]
Italy: Joab ben Jehiel Beth-El, fourteenth century
The Vatican (Ebr. no. 462)

98

Lungoburgo was a thirteenth-century Christian physician. His *Chirurgia magna*, a lengthy and famous treatise on surgery, was composed in 1252 and translated into Hebrew by Hillel ben Samuel. The work, which combines the Arabic surgical traditions available in Latin with Bruno's own ideas, was also issued in an abridged edition.

Jacob ben Asher of Toledo (ca. 1270–1340)

ארבעה טורים
[Arba'ah Turim]
Soncino: Solomon ben Moses Soncino, 1490 (?)
Columbia University, New York

99

The only complete medieval codification of Jewish law was *Maimonides' *Mishneh Torah*, but the author's omission of all discussion of the sources on which his decisions were based and the availability of other medieval approaches to many issues suggested the need for a legal digest that would summarize the extant positions and arguments. This was accomplished by Jacob ben Asher's seminal work, *Arba'ah Turim*.

The *Arba'ah Turim* is a topical presentation of medieval discussions and decisions about all practical aspects of Jewish religious law, and it therefore differs from Maimonides' effort in two important ways: it is not a code, and it omits those aspects of Jewish law not in use (e.g., those applicable only to the Temple rituals and the messianic era). As suggested by its name, which derives from the four rows of stones on the priest's breastplate (Ex. 28:17), the *Arba'ah Turim* contains four sections. These deal with, respectively, the liturgy, and Sabbath and holiday observances; other matters of ritual law, including dietary restrictions, ritual slaughter, circumcision, Torah study, and mourning; marriage, divorce, and related issues; and civil law. As both a summary of many different legal positions and an attempt at choosing among them, the *Arba'ah Turim* stimulated adjudications and commentaries by many later writers. Chapter 336 of *Yoreh De'ah* presents the rationale for allowing medical treatment, the basis for assuming the right to avert the presumably divinely ordained impact of illness, and related matters of medical ethics.

Joseph ben Ephraim Caro of Turkey and Safed (1488–1575)

שולחן ערוך מיטור יורה דעה הנקרא בית יוסף
[Shulhan Arukh MiTur Yoreh De'ah HaNiqra' Beit Yosef]
Venice: Giovanni Grifio (Grypho), Adar, 1567
Jewish Public Library, Montreal

100

Born in the Iberian Peninsula but exiled to Turkey, Caro became a major leader in the re-established community of Safed. A mystic and the recipient of inspirational teachings

LITTERAE
S. D. N. D. GREGORII PAPAE XIII.

Innouationis constitutionum Pauli Quarti, & Pij Quinti, contra Medicos Hebreos, Et illarum extensionis ad eos qui Medicos Hebreos, vel infideles ad Christianorum curam vocant, admittunt, vel eisdem medendi licentiam concedunt.

GREGORIVS PAPA XIII.
Ad perpetuam rei Memoriam.

LIAS Piæ me. Paulus Papa IIII. prædecessor noster edita perpetua constitutione inter alia sanciuit, ne Medici Iudæi etiam vocati, & rogati ad Christiano um ægrotantium curam accedere, aut illi interesse possent: quam constitutionem postea P us Papa V. etiam prædecessor noster per suas litteras approbauit, innouauit, & confirmauit, & robur perpetuæ firmitatis obtinere decreuit, ac voluit, & sub interminatione diuini iudicij præcepit, & mandauit, omnia in eadem constitutione contenta in pcsterum firmiter observari non solum in Terris, & Dominijs Sanctæ Romanæ Ecclesiæ subiectis, sed etiam vbique locorum. Qua tamen nobis non sine magna animi nostri molestia annotuit, a minime observari, sed multos adhuc ex Christianis hominibus esse, qui dum suos corporum languores illicitis medijs, & præcipue Iudæorum, ac aliorum infidelium opera sanari cupiunt, veræ salutis animarum suarum, & corporum simul immemores siunt, & (quod valde dolendum est) in damnationis æternæ maximum sæpe discrimen incidunt, medicis Iudæis, & infidelibus huiusmodi ad ipsorum curationem vocatis, & adhibitis: vnde fit, vt & Iudæis, ac alijs infidelibus magna detur delinquendi occasio, & simul salutare præceptum negligatur ab Innocentio Papa III. similiter prædecessore nostro in Concilio generali quondam emissum, & deinde à prædicto Pio V. innouatum, quod omnes medici cum ad infirmos in lecto iacentes vocati essent, ipsos ante omnia monerent, vt idoneo consessori omnia peccata sua iuxta ritum Sanctæ Romanæ Ecclesiæ confiterentur, neque tertio die vlterius eos visitarent, nisi longius tempus infirmo confessor ob aliquam rationabilem causam, super quo eius conscientia onerabatur, concessisset, & eis per fidem confessoris in scriptis factam constaret, quòd infirmi peccata sua confessi fuissent. Idcirco nos tam Iudæos, qui aduersus mandata huiusmodi apostolica committere audent, quam Christianos qui illos ad se accersunt. vel medendi licentiam concedunt, & viam ad delinquendum eisdem aperiunt, coercere volentes, supradictas prædecessorum nostrorum constitutiones auctoritate apostolica tenore præsentium approbamus, confirmamus, & innouamus, ac inuiolabiliter observari mandamus, itque hac nostra in perpetuum valitura constitutione, eisdem constitutionibus, & præceptis pro firmiori il'orum observatione addentes, vniuersis vtriusque sexus Christifidelibus districte inhibemus, & interdicimus, ne posthac Iudæos, vel alios infideles ad ipsorum Christ anorum ægrotantium, & infirmorum curam vocent, seu admittant, aut vocari, admittue faciant, concedant, vel permittant. Mandantes propterea omnibus, & singulis venerabilibus fratribus nostris Patriarchis, Primatibus, Archiepiscopis, & Episcopis, necnon dilectis filijs alijs locorum ordinarijs, & quibusuis Parochis, alijsue animarum curam habentibus, & exercentibus, sub indignationis nostræ, ac alijs arbitrio nostro infligendis pœnis, vt præsentes nostras litteras in suis ecclesijs, quæ in illis ciuitatibus, vel diœcesibus constitutæ sunt, in quibus Hebræi, vel alij infideles moram trahunt, quamprimum ad eos perlatæ fuerint, & deinde singulis annis initio quadragesimalis ieiunij publicent, aut publicari faciant, & quod si quis post earum publicationem etiam quomodolibet ex imptus ac cuiuscunque status, gradus, ordinis, conditionis, & præeminentiæ existens, aduersus illa facere ausus fuerit, sacramenta ei ecclesiastica nullatenus ministrentur, nec etiam à regularibus exemptis: & sic decedens, ecclesiastica careat sepultura, quæ quidem omnia parochi ægrotantibus significare apto tempore non omittant, præsertim cum uduum, vel infidelem medicum ab eis admissum esse cognouerint, & aliàs ipsi locorum ordinarij contra huius mandati transgressores debita animaduersione procedant: Iudæosque ipsos nihilominus iuxta Pauli, & Pij Pontificum prædictorum; literas contra illos editas pro earum transgressione puniant. Non obstantibus constitutionibus, & ordinationibus apostolicis, ac omnibus illis, quæ iidem Paulus & Pius in suis litteris prædictis voluerunt non obstare, priuilegijs quoque indultis, & litteris apostolicis quibusuis personis etiam regularibus, priuilegiatis, & exemptis, eorumque ordinibus, & congregationibus sub quibuscunque tenor bus, & formis, etiam Mari Magno, seu bulla aurea nuncupata, ac cum quibusuis clauulis, & decretis in genere vel in specie, ac al às in contrarium quomodolibet concessis, approbatis, & innouatis, quibus omnibus, etiam si de illis, eorumque totis tenor bus, & formis speciali, specifica, expressa, & indiuidua mentio, seu quæuis alia expressio habenda, aut aliqua alia exquisita forma ad hoc seruanda foret, illorum tenores, ac si de verbo ad verbum nihil penitus omisso, & forma in illis tradita observata inserti forent, præsentibus pro sufficienter expressis habenteis, illis aliàs in suo robore permansuris, hac vice duntaxat specialiter, & expresse derogamus, contrarijs quibuscunque. Seu si aliquibus communiter vel diuisim ab apostolica sit sede indultum, quod interdici, suspendi, vel excommunicari non possint per litteras apostolicas non facien es plenam, & expressam, ac de verbo ad verbum de indulto huiusmodi mentionem. Et quia difficile foret easdem præsentes ad singula quæque loci deferri, volumus, & declaramus, quod earum transumptis, etiam impressis, & manu alicuius No arij publici subscriptis, ac sigillo alicuius personæ in dignitate ecclesiastica constitutæ muni is, eadem prorsus fides vbicunque habeatur, quæ præsentibus haberetur, si forent exhibitæ, vel ostensæ. Dat. Romæ apud S. Petrum sub Annulo Piscatoris, die xxx. Martij, M.D.LXXXI. Pontificatus nostri Anno Nono.

Cæ. Glorierius.

Anno a natiuitate Domini Millesimo Quingentesimo Octuagesimo primo, Indictione Nona, die vero Quinta Mensis Aprilis, Pontificatus Sanctissimi in Christo Patris & Domini Nostri D. Gregorij diuina prouidentia Papæ XIII. Anno Nono, retroscriptæ litteræ apostolicæ affixæ, & publicatæ fuerunt ad valuas Cancellariæ Apostolicæ, & in acie Campi Floræ, per me Petrum Aloysium Gaytam eiusdem Sanctissimi Domini Nostri Papæ Curs.

Petrus Stoch Magister Curs.

IN ROMA, Per gli Heredi d'Antonio Blado Stampatori Camerali. M.D.LXXXIIII.

Papal letter of Gregory XIII that banned Christian use of Jewish physicians, Rome, 1583. McGill University, Montreal, Osler Library (Cat. no. 101)

from a personal spirit (*maggid*) believed to be the personification of the Mishnah, he wrote a very important commentary on Jacob ben Asher's *Arba'ah Turim*, *Beit Yosef*. Not satisfied with merely annotating and augmenting Jacob Ben Asher's digest, he also condensed these legal teachings into a code, the *Shulhan Arukh*, which followed the topical organization and literary structure of the *Arba'ah Turim*, while presenting each legal decision in an independent and easily identified paragraph. In the section on visiting the sick, Caro codified some of the legal and ethical issues related to the practice of medicine.

The *Shulhan Arukh* has been printed in many editions, almost all of which contain glosses by Moses *Isserles – which bring its Sefardic approach into line with Ashkenazic legal practice and tradition – and commentaries. Indeed, its acceptance as the guiding work of Jewish religious practice took time and was not completed until it was supplemented by a series of detailed and learned commentaries during the subsequent centuries. The first two editions were issued in Venice in 1565 and 1567 without Isserles' glosses, which were added in Cracow in 1568. The Vilna edition, accompanied by over a dozen commentaries on every page, has one of the most complex layouts of any Hebrew text ever published.

Pope Gregory XIII (1502–85)
Litterae S.D.N.D. Gregorii Papae XIII
Rome: Heredi d'Antonio Blado, 1583
McGill University, Montreal (Osler Library no. 2827)

101

This papal letter, initially posted on 30 March 1581, renewed the decrees of Popes Paul *IV* (died 1559) and Pius *V* (died 1572). The former prohibited Jewish physicians and other infidels from treating sick or infirm Christians, even if they were summoned by the Christians themselves. The latter added that the decree was to be observed eternally, everywhere, even in lands not under the church's dominion.

These strictures were enacted in order to ensure the spiritual well-being of the patients and to guarantee that their lives would not be saved through illicit means, that is, medical treatment not in keeping with Christian beliefs, including proper confession. The penalty for failing to publish the decree locally was papal indignation and other penalties to be inflicted at papal discretion. The penalty for violating the decree – which is directed both at the physicians and at those who summon or admit them – is withholding of the sacraments (hardly a serious threat to the physicians) and, in the case of death, denial of a church burial. All exemptions, indulgences, and special permits in force (a popular technique used to circumvent such decrees) were hereby revoked.

This type of letter points to the church's frustration with the professional success of an entire cadre of Jewish physicians, a number of whose works are mentioned in this chapter. One should not forget that some popes, including Paul *III* (1534–49), maintained Jewish personal physicians. David *de Pomis, for one, was affected by this decree and responded to it with a published plea to recognize the potential humanitarian contribution of Jewish physicians, but to no avail. It is generally agreed that the decree hastened the decline of Jewish medical practice in Italy, which had been quite illustrious.

Gregory *XIII* is perhaps best known for his calendar reforms. His attitude toward Jews vacillated during his thirteen-year papacy (1572–85), and his actions ranged from benign or even helpful to quite hostile. Another decree, issued on 1 June 1581, gave the Inquisition license to handle matters of Jewish blasphemies.

A full, handwritten English translation of the Latin original has been appended to this copy of the letter.

David [ben Isaac] de Pomis of Venice (1525–1593)
Brevi Discorsi et Eficacissimi Ricordi per Liberare Ogni Citta Oppressa dal Mal Contagioso
Venice: 1577
Biblioteca Palatina, Parma

102

David de Pomis was born in Spoleto in 1525 and, like many other Italians, traced his ancestry back to a family exiled to Italy in the time of Titus. After studying medicine, he served as a physican and rabbi in the community of Magliano, near Rome. During his lifetime, papal attitudes toward Jewish physicians vacillated from relatively benign to

עוד ייעזר העניין גם יתברך לפרנס עשר ידיה
חתוך שאבן פעם העראה עור יין לערת לא
יחתוך ושם בליפות האותיות **טספץ** וחותן
עיה והאור דות **סומפר** כו ויגהל עוור
אלא לחתון בליפין טך יה יין אלבע עטנ שרא
זרה יעני יעט יעי אוא ולחתון הסי אל אושוה

לאשה היולעת דם פתיחה הדה קשה על חנור חדש
אל חשות וחלו אלה פטים ועלה יוער דם מחנור
החנור זה יקרא יה יה יה מיכאל גבריא
ויסעו ויט וטי עד ל הפסיק ואחר השמות תחת
אותם כאנה

עץ לאשה לכד לוחש באונה בא לא בא א
ירכי יה ובני אדם את ירפאים האוסרים
עיר עירי עד היוד קה על

עץ **אהו** כאונה השמות יה **אורטם**
בורגם אל ארום כה

ארחת
כריד גג הלחות עד הכוו אז עלעי כפי ויתה
בתואה וצריך לחיד ויהיו וו ע" ה **ארכתתנינא**
דכתאינום כתאונום ועכום
ישוע עין קין ע

harshly negative, and de Pomis, who had received papal permission to practise medicine, moved to Venice, where he could work unencumbered.

He wrote several important medical works. *Brevi Discorsi* deals with plague; *Enarratio Brevis de Senum Affectibus Praecavendis atque Curandis* discusses the diseases of old age. *De Medico Hebraeo Enarratio Apologica*, occasioned by *Gregory XIII's papal bull of 1581, contains a call for the physician's integrity and commitment to healing anyone in need of his services, and a lengthy, impassioned, and generally ignored plea for recognition of the contribution Jewish physicians can make to the general welfare.

In addition to his Italian translations of Job, Daniel, and Ecclesiastes (of which only the last has been published), and a Hebrew-Latin-Italian dictionary (*Zemah David*, Venice, 1587), de Pomis mentions a work he wrote on the battering ram.

Rodrigo de Castro Lusitanus of Lisbon and Hamburg (1546–1627)
De Universa Muliebrium Medicina ... Volume 1
Hamburg: Phillip de Ohr, 1603
Hebrew Union College, Cincinnati

|103|

Medicus Politicus
Hamburg: Frobeniano, 1614
Hebrew Union College, Cincinnati

|104|

One of the earliest and most important members of a distinguished family of Marrano physicians, de Castro practised in Lisbon. He was invited to take the place of *da Orta in Goa, but declined. His talents were highly respected by Zacutus *Lusitanus, who described him in superlatives.

De Castro moved to Antwerp and later to Hamburg, where he was a founding member of the Jewish community. Though he and his co-religionists were not made welcome, his search for greater religious freedom served the interests of his new neighbours. He contributed much to the battle against the plague of 1596, which is discussed at length in *Tractatus brevis de natura et causis pestis quae hoc anno MDXCVI Hamburgensem civitatem affligit*. It, *Medicus Politicus*, and *De Universa Muliebrium Medicina* (which contained the latter in some editions) were all published many times.

Abraham Zacutus Lusitanus of Lisbon and Amsterdam (1575–1642)
Opera Omnia, Volume 1
Lyon: Ioannis-Antonii Hvgvtan, 1642
Hebrew Union College, Cincinnati

|105|

Tratado Sobre Medecina que fez o Doritor Zacuto para seu filho levar consigno quando se foy para o Brazil
B. Godines, 5450 (1690)
Columbia University, New York (X610 Sa2)

|106|

Abraham Zacutus Lusitanus, generally known as Zacutus Lusitanus, was a descendant of Abraham *Zacuto and the scion of a very well-known and learned family. He was a Marrano who studied medicine at Coimbra and Salamanca, and practised it in Lisbon for thirty years. At the age of fifty, he moved to Amsterdam, where he became a close friend of Menasseh ben Israel.

Lusitanus maintained an abiding reputation as a leading medical figure, and a careful examination of his work reveals his deep concern with every facet of the profession. He was particularly involved in the study of diseases, and made important contributions to the knowledge of plague, diphtheria, syphillis, and blackwater fever; he also made the first proper identification of several diseases and ailments.

Tratado Sobre Medecina is a small volume composed by Lusitanus for use by his son during his travels in Brazil. It is a handbook of medicinal information to be used in the primitive circumstances of the New World in which he was about to find himself.

Shabbetai ben Meir HaKohen [Shakh] of Lithuania and Poland (1621–1662)

גבורת אנשים

[Gevurat Anashim]
Dessau: Moses ben Simhah Bunem, 1697
National Library of Canada, Ottawa (Jacob M. Lowy Collection)

107

Except in limited and sometimes criticized ascetic circles, Jewish thinkers have considered sexual expression and reproduction natural and necessary functions. To be sure, these were associated with, and sanctified by, marriage, but the recognized need for a fulfilled sex life (for both men and women) dominates many talmudic discussions of marriage and related legal concerns.

In this context, the implications of sexual impotence take on added religious importance, and the Talmud discusses fully the status of a marriage that cannot be consummated or cannot provide adequately for the wife's needs due to male impotence. These matters are developed through the normal channels of halakhic discourse and are codified in the *Mishneh Torah*, *Arbaʿah Turim*, *Shulhan Arukh*, and *Levush*, as well as in specialized halakhic works. They are also treated in many of the thousands of volumes of responsa composed by rabbis since early medieval times.

Shabbetai ben Meir HaKohen, better known in rabbinic circles as the Shakh, was a leading Lithuanian rabbi and author of one of the most important commentaries on *Yoreh Deʿah*, the second volume of Joseph Caro's *Shulhan Arukh*, *Siftei Kohen* (one source of his acronym). He also wrote commentaries on the *Yoreh Deʿah* section of Jacob ben Asher's *Arbaʿah Turim* and on the fourth part of the *Shulhan Arukh*, *Hoshen Mishpat*. *Gevurat Anashim* is based on chapter 154 of *Even HaEzer* (the third section of the *Shulhan Arukh*), which deals with the legal grounds for divorce, including claims of impotence brought by the wife.

Abraham Wallich (died 1693) and Judah Leib Wallich (1688–1735) of Frankfort

ספר דמיון הרפואות

[Sefer Dimyon HaRefu'ot] or *[Harmonia Wallichia Medica]*
Frankfort on Main: Johanne Wust, 1700
Hebrew Union College, Cincinnati

108

The Wallichs, father and son, were members of a German family that had for generations included physicians. *Sefer Dimyon HaRefu'ot* is a small Hebrew medical manual containing information on a range of topics. Based on the collected papers of Abraham, but published posthumously and augmented by his son, the book contains much material derived from Jacob ben Isaac Zahalon's *Otzar HaHayyim*, published in Venice in 1683. It has been suggested that Judah Leib found a copy of parts of the *Otzar HaHayyim* made by his father for his own use and, thinking it to be original, published it; but it is not impossible that the material was simply borrowed and issued under a new name. *Dimyon HaRefu'ot* combines what we would call medicine and psychology, as one of its goals is a comparison of the ills of the body and those of the soul.

Judah Leib's poetic introduction to *Dimyon HaRefu'ot* contains a strongly worded attack against uneducated women who compete with trained physicians and do more harm than good. He shared this attitude with Tobias *Cohn, who expressed equally harsh sentiments in the poetic parts of his *Maʿase Tuviah*.

Abraham ben Solomon Hamburg of Nansich (Nanzig)

עלה תרופה

[Aleh Terufah]
London: Alexander bar Judah, 1785
Jews College, London

109

When faced with various types of problems or challenges, primitive societies often responded by adopting rather than avoiding them. Thus, attack by wild beasts might lead to wearing a part of the animal's hide or a tooth; antidotes for poisons were often sought in the poison itself; and those who overindulged in alcohol often took an additional drink during the recovery period. Whether done to instigate the problem at a lesser level than might occur otherwise, to fortify oneself symbolically by identifying with the enemy, or because the procedure actually improved the situation, this type of reasoning seems to lie behind the practice in some places of inserting smallpox blisters or other infectious sores under the skin or into the nostrils of unaffected people, which served as a form of vaccination. The formal process of vaccination against smallpox was performed on two members of the British royal family in 1722, and it temporarily became quite popular until fears of the procedure grew, and it was then banned in a number of places.

Halakhah governs all practical aspects of daily life and devotes attention to matters of medical practice and ethics. The importance of medicine to the Jewish ethos and the high number of Jews who had practised it since medieval times served to keep medical issues before the Jewish courts. One therefore finds that rabbis discussed the implications of many medical discoveries and procedures, particularly when they seemed to run counter to the natural assumptions of good health. Under the circumstances, it should not be surprising that vaccination, which required infecting the patient with a potentially fatal disease, attracted their attention.

The author of *Aleh Terufah* first identified smallpox by its Hebrew, German, English, French, and Italian names, and described in detail the vaccination process. He then explained the cause of the disease as presented in Jacob Zahalon's *Otzar HaHayyim*, which assumed that its potential was already in the body at the time of birth.

The permissibility of vaccination rests on the resolution of a conflict between two halakhic values: the paramount importance of saving a life and the prohibition on unnecessarily placing oneself in danger. In classic rabbinic fashion, the author discussed appropriate talmudic sources, found support in various biblical verses, dismissed arguments that temporary illness might interfere with observance of ritual commandments and that rare deaths from the procedure reflect negatively on it, and ruled in its favour.

Despite the clear explanations at the beginning and end of the essay noting the connection between the title and his name, the author's identity has been questioned. The title page suggests that it was written by "Abraham ben Solomon Hamburg, who also bears the name Abraham Nansich (Nanzig)." According to Ben Jacob (*Otzar HaSefarim*, p. 441), the same material appeared in *HaMeasef* in 1785 and was reissued with some novellae on a few talmudic passages in London that same year. Friedberg (*Beit Eked Sefarim*, p. 796) attributes the work to Rabbi David Tevele Schiff (rabbi of the Great Synagogue of London, died 1792) and an added *derush* for the completion of a tractate of the Talmud to Nansich.

DISEASES OF THE EYES

The eyes rank among the body's most important organs, and their accessibility renders them both highly vulnerable and observable. Together, these factors contributed to the eyes being one of the most frequently discussed medical topics. Many general works include sections on eyes, eye diseases, and treatments for the eyes, and specialized ophthamological treatises were also written.

The apocryphal book of Tobit, which tells of blindness caused by a bird defecating in Tobias's eyes and the treatment that cured it (a fish gall), was a favourite topic for discussion; but the Middle East, in general – with its poverty, lack of water and poor level of hygiene – was famous for eye diseases. One talmudic figure taught that an unwashed hand that touches an eye deserves to be cut off, because it causes blindness. And *Maimonides stressed that washing one's hand for hygienic reasons after a meal was at least as essential as washing them for ritual purposes before it (*Mishneh Torah, Hilkhot Berakhot*, chapter 6).

Anonymous
Fragment of a lengthy Assyrian medical work
Cuneiform, eighth or early seventh century, BCE
McGill University, Montreal (Osler Library, no. 53)

110

The oldest text in the exhibition is a cuneiform tablet from biblical times that contains a piece of the thirty-second tablet of a lengthy medical or scientific work. It contains a series of prescriptions for treating eye problems, though several applications to other parts of the body are also discussed. Akkadian, including the northern (Assyrian) and southern (Babylonian) dialects, was the Middle Eastern *lingua franca* of biblical times. It also served as the language of scientific compositions that demonstrate pre-Hellenic mastery of various aspects of mathematics and astronomy. Akkadian literature is replete with omen texts that offer historical predictions in conjunction with the appearance of

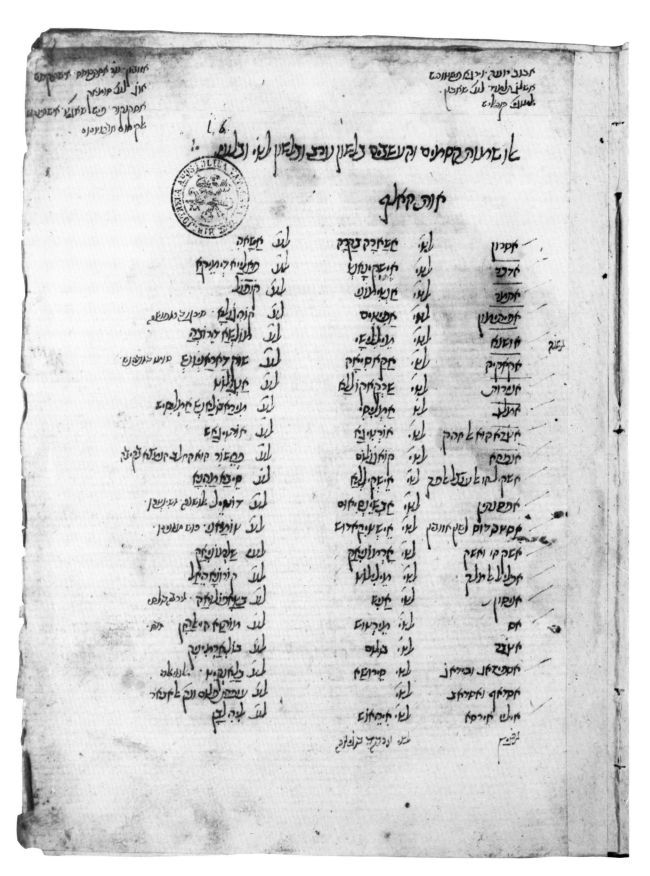

A trilingual list of plants and medicinal herbs. *Shemot HaAsabim VeHaSamim*. The Vatican, Ebr. #356 (Cat. no. 114)

anomalous anatomical changes in the internal organs of sheep. The prescriptions found here share the orientation of these omen texts. They contain combinations of potions to be applied to the eye, and often conclude with an omen-type formula, "he will live."

Paleographic considerations moved S.H. Langdon to date the tablet in the reign of either Sargon or Sennacherib. This means that the tablet, a copy of an earlier Babylonian work, was produced around the time of King Hezekiah of Judea and Isaiah, one of the most prominent biblical prophets; and it may reflect the type of treatments administered during that period. It is striking evidence of the long-standing concern for the treatment of ophthamological ailments in the Middle East.

Among the curatives prescribed are swamp melon, oil, dog's tongue (i.e., hedge mustard), scorpion blood obtained by pulling out its tongue and cutting off its head, and serpent's blood. Each of these was applied to the eye, often while the patient was being encouraged to drink good beer.

The owner of the tablet, Kisir-Ashur, son of Nabu-tuklat-su, was the grandson of a magician in the temple of Ashur and the great-grandson of a libation priest. He thus had a noble lineage and, like the biblical priests, was entrusted with the medical knowledge associated with his office.

Ammar ibn Ali al-Muʿsuli Abul Kasim (eleventh century)

ספר ברפאות הצין
[Sefer BeRefu'ot HaAyyin]
Rabbinic script, fifteenth century
Biblioteca Palatina, Parma (no. 3047)

111

Rabbinic script, fifteenth century
Biblioteca Palatina, Parma (no. 3590)

112

Abul Kasim (Abu Al-Kassem) or Abulcassem, as he is better known in the West, was an eleventh-century Egyptian physician. His Arabic work on ophthamology was translated into Latin, and also into Hebrew by Nathan HaMeʿati. Nathan, a translator of *Avicenna's Canon, was one of the best known medieval translators of Arabic scientific works into Hebrew.

Marcus (Mordecai) Elieser Bloch of Berlin (1723–99)
Medicinische Bemerkungen
Berlin: C.F. Himburg, 1774
McGill University, Montreal (Osler Library)

113

*Bloch, an important zoologist, also contributed to the medical profession. His books are regularly accompanied by interesting, and often beautiful, illustrations; his text on the eyes is no exception.

MEDICINES

The preparation of appropriate and effective medicines was one of the physician's most important tasks. To this end, many generations of medical practitioners – including priests, witch doctors, community elders, quacks, and trained doctors – have collected and produced vast works describing potions, salves, herbs, and chemicals designed to improve the condition of sufferers. Many general medical works include lengthy sections or whole volumes on this subject, and many individual treatises also were devoted it.

These pharmacologies usually take the form of lists, often multilingual, that aid in the identification of the *materia medica* in several of the languages available at a given time and location. Some examples are fully illustrated, providing an additional aesthetic quality as well as furthering the precise identification of the plant or plant part under discussion.

Title page of Garcia da Orta's *Coloquios dos simples,* Goa, 1563. The Wellcome Institute Library, London (Cat. no. 117)

Bilingual prescriptions for treating red and white phlegm. Meir ben Solomon Alguadez, *Sefer Meiqitz Nirdamim.* Montefiore Endowment, Jews College, London, MS no. 444 (Cat. no. 115)

Anonymous

שמות העשבים והסמים
[Shemot HaAsabim VeHaSamim]
Rabbinic script, fifteenth century
The Vatican (Ebr. no. 356)

This thirty-seven-folio herbal list is trilingual, with the items listed in Arabic, Hebrew, and Latin, all written in Hebrew letters. The manuscript also contains a list of days on which bloodletting should be done, and several astronomical works by Jacob ben David, including *Sefer HaItztagnut*, astronomical tables, and a work on the calendar.

Meir ben Solomon Alguadez of Castile (died 1410)

ספר מקיץ נרדמים
[Sefer Meiqitz Nirdamim]
Spanish cursive, sixteenth century
Montefiore Endowment, Jews College, London (MS no. 444)

Alguadez was a skilled fourteenth-century doctor who served as court physician and chief rabbi of Castile by appointment of the king. He also translated Aristotle's *Ethics* into Hebrew. His collection of prescriptions was written in Spanish, and was translated by Joseph ben Joshua HaKohen (1496–1578), author of several important historical works, including *Emeq HaBakhah*.

The Hebrew version of Alguadez's prescriptions is found in *Sefer Meiqitz Nirdamim* with a Latin version, which greatly simplified identification of the ingredients. It also contains a larger section of additional prescriptions by Joseph HaKohen, plus other collations of prescriptions and related additions.

Garcia da Orta of Lisbon and Goa (ca. 1500–68)
Aromatum, et simplicium aliquot medicamentorum apud Indos nascientium historia ... nunc ver primum Latina facta ...
Antwerp: C. Plantin, 1567
Wellcome Institute Library, London

Coloquios dos simples, e drogas he cousas medicinais da India, e assi dalgunas frutas achadas nella onde se tratam algunas cousas tocantes a medicina, practica, e outras cousas boas ...
Goa: J. de Endem, 1573
Wellcome Insitute Library, London

Da Orta was born in Portugal, studied medicine at Salamanca and Alcala, and later taught logic at Lisbon. In 1534 he left for India, where he served as physician to many important political leaders and church figures. His *Aromatum et simplicium ...* a study of drugs and medicinal plants, appeared in at least six Latin editions between 1574 and 1611, and also in Spanish, Italian, and French.

In Goa, da Orta spent many years studying plants and medicinal herbs, whose properties he analyzed thoroughly and rigorously. There he wrote, in Portuguese, *Coloquios dos simples, e drogas he cousas medicinais da India ...* It was published with the approval of the Inquisitional authorities, and its importance is best indicated by the impact it had on the study and use of medicinal herbs, and by its subsequent issue in Italian, French, Spanish, and English versions.

Upon investigation, the Inquisition determined that several of da Orta's relatives, including his sister, were Marranos, and further searching implicated the famous physician and botanist himself. By this time, da Orta had died, but, having been cheated of their live victim and fooled by so prominent a medical and scientific personality, the inquisitors had his body exhumed and burnt at the stake.

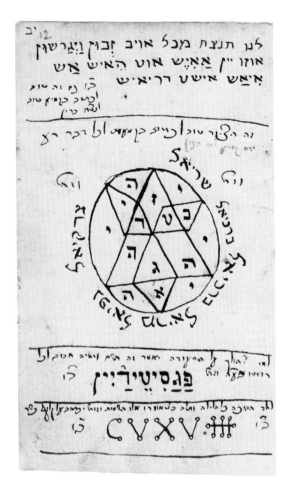

Hebrew prescriptions and models for charms and amulets. The Wellcome Institute Library, London, MS Heb. A 19 (Cat. no. 118)

Yiddish prescriptions. *Liquttei HaRefu'ot*. The Vatican, Ebr. no. 371 (Cat. no. 119)

Anonymous
Untitled Yiddish prescriptions and miscellanies
German cursive, seventeenth or eighteenth century
Wellcome Institute Library, London (Hebrew no. 19)

118

Anonymous

לקוטי הרפואות
[Liquttei HaRefu'ot]
German cursive, sixteenth century
The Vatican (Ebr. no. 371)

119

The nineteenth and twentieth centuries saw the development of a vast Yiddish literature that included scientific and mathematical works, but the documents available from earlier centuries were generally devoted to other interests. Among them are numerous small volumes containing mixtures of calendrical information, practical advice for personal problems, folk cures, charms, and remedies. Wellcome Institute manuscript no. 19 contains 206 prescriptions, magical charms, instructions for folk cures and amulets, and nursery rhymes. Vatican Ebr. no. 371 contains similar types of material.

MAGIC AND MEDICINE

The aspects of human culture that contemporary wisdom designates as magic, religion, and science were not clearly differentiated in pre-modern times. Though many societies outlawed magic and supported religion (or, at least, their preferred religion), it was often impossible to make distinctions between them. As late as the seventeenth century it was fashionable to integrate them, a tendency that can be observed even today. What we recognize as scientific discovery was often considered revelation several thousand years ago; even in later centuries, when discoveries were recognized as such, the search to find meaning in them often coloured the ways in which they were perceived and used.

Since members of pre-modern societies often attributed powers or personalties to what we see as inanimate objects or natural forces, their responses to them were very different. Disease, with all its accompanying confusion and potential destruction, was poorly understood, and plagues ravaged the cities with terrifying regularity. Until the causes of particular diseases were understood in ways that allowed them to be both prevented and cured – and the ability to do this varied greatly in different times and places – magical and scientific means of coping with them were not always distinguished, and sometimes supposed cures actually contributed to the spread of disease. On the other hand, procedures followed during such crises often made at least two positive contributions. In some cases, what appeared as magical praxis actually anticipated scientifically defensible cures; in others, it provided essential psychological support in the battle against the invisible hostile forces that were arrayed against mankind.

The primitive anticipation of medically valid cures belongs to the realm of folk medicine. Psychological aid was provided by many rituals and sacred objects, including amulets and talismans, which have no inherent therapeutic value. Over the centuries, many different religious and cultural groups – including many Jewish ones – relied on the power of amulets and charms to prevent and cure physical and psychological ills.

Aramaic, the *lingua franca* of the Middle East from before the rise of the Persian Empire to after the rise of Islam, was used by Christians, Jews, and virtually every other group that made its home between India and Europe. It was the language of Egyptian papyri, Syrian inscriptions, Mesopotamian tablets, Jewish prayers, some of the Dead Sea Scrolls, the eastern patristic literature, gnostic liturgies, the Talmud, many midrashim, New Testament figures, including Jesus, and many more. It was used by many peoples, of many faiths, and by both ordinary folk and scholars.

The culture of this period included belief in demons and in the efficacy of magical incantations for coping with them. Aside from preferences for particular demons and/or angels and the ritualistic backgrounds from which they and their antidotes were drawn, one notices little difference between the ways various religious groups responded to them. They perceived demons as real problems and attacked them with universally recognized forces, just as we generally treat illness in similar ways, regardless or our religious affiliations. The Aramaic magic bowls were one of the means by which they did this.

Each bowl contains a circular inscription, usually written around the inside, but occasionally outside, as well. Some are illustrated. The text, written individually for each

Aramaic bowl with incantation and drawing. University Museum, Philadelphia, CBS 2923 (Cat. no. 122)

patron, is intended to protect the people mentioned in it, and their possessions, from demons, evil forces, or illness. Jews, Christians, and Mandaeans produced these bowls in their distinctive Aramaic dialects and scripts. The Jewish bowls often cite the Bible, and occasionally they refer to known rabbinic figures or to rabbinic lore and Bible interpretations.

Pabak bar Kufithai
Aramaic prophylactic bowl
Nippur: Pabak bar Kufithai, seventh century (?)
University Museum, Philadelphia (CBS 2945)

120

This bowl was prepared to protect Abuna bar Geribta and Ibba bar Zawithai from a series of evil forces, and its writer drew his power from the garment of Hermes and the Creator of heaven and earth. He threatened the destructive forces with the curses of the Leviathan and Sodom and Gemorrah.

The circular format of the text is normal for magic bowls, but the presence of an illustration is unusual, though not unique. The figures are not labelled, and their identities and purposes are not certain. J.A. Montgomery, who first published the Nippur bowls, suggested that the figure with the bound feet is a demon and that the other figure is the sorcerer.

Anonymous
Aramaic prophylactic bowl
Nippur: seventh century (?)
University Museum, Philadelphia (CBS 2963)

121

This bowl is directed at protecting the household of Ardoi bar Hormizduch from the angel of death, who takes husbands from their wives, wives from their children, and children from their parents. Its medical focus is clear from the first line, which begins with an appeal to the "master of healing." The bowl concludes, "May the Lord rebuke you, Satan; may the Lord who has chosen Jerusalem rebuke you. Is this not a brand plucked from the fire? [Zechariah 3:2] Amen. Amen." This verse's strongly worded reference to God's attacking Satan led to its being the most popular biblical text in the magic-bowl literature.

The illustration was interpreted by Montgomery as an armed and shackled demon. Given the contents of the bowl, one might conclude that it is a representation of the angel of death.

Anonymous
Aramaic prophylactic bowl
Nippur: seventh century (?)
University Museum, Philadelphia (CBS 2923)

122

This bowl was prepared to protect Pabak bar Kufithai, mentioned in the above description, and his household. The demonic forces are to be controlled by the powers of Enoch, the seven planets, and the twelve signs of the zodiac. The sketch seems to depict a sorcerer waving a branch in an attempt to ward off the demons, and bears a resemblance to a picture in Oslo Magical Papyrus No. 1, reproduced by Doresse.

Anonymous
Aramaic prophylactic bowl
Mesopotamia, seventh century (?)
Royal Ontario Museum, Toronto (907.1.1)

123

This bowl was written to protect the household of Babai bar Mahlafta from five angels who afflict it. It is sealed with the seal of El Shaddai.

Aramaic bowl with incantation and drawing. University Museum, Philadelphia, CBS 2945 (Cat. no. 120)

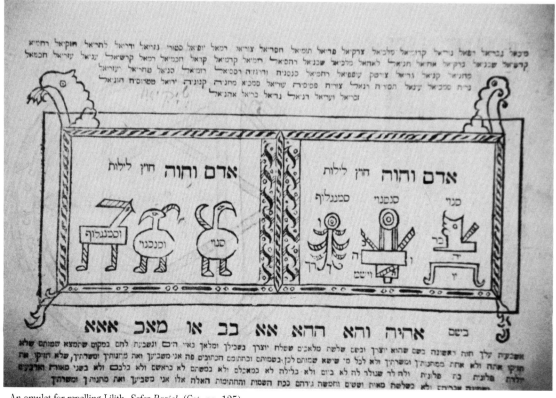

An amulet for repelling Lilith. *Sefer Raziel.* (Cat. no. 125)

Anonymous
Aramaic prophylactic bowl
Mesopotamia: seventh century (?)
Royal Ontario Museum, Toronto (907.1.2)

This poorly preserved text was written to protect the health of one Jannai bar Zacut. The illustration in the centre of the bowl is also in very poor condition, but seems to contains a pair of bound legs. If so, it probably represents the same type of demonic figures found in other bowls.

Anonymous

ספר רזיאל

[Sefer Raziel]
Amsterdam: Moses ben Abraham Mendes Coutinho, 1701
Annenberg Research Institute, Philadelphia

One of the most feared demons was Lilith. According to ancient Jewish lore, Lilith was Adam's first wife, who rebelled against him and his sexual preferences and became the prototype of the succubus. Anxious to ensure the propagation of demons, Lilith, some people believed, seduced men during their sleep and enticed them to sire infinite numbers of demon offspring. Her other passion was causing the deaths of human children.

Whether she was a vestige of ancient Mesopotamian mythology or the personified cause of primitive societies' high infant mortality rates – perhaps the mysterious phenomenon we have come to call crib death – or both, Lilith was a feared foe who could be thwarted only through the proper procedures. Many amulets placed on or near children had the purpose of shielding them from her; one of the most famous is found in *Sefer Raziel*. *Sefer Raziel*, reputed to protect from fires the dwelling in which a copy was kept, was one of the most famous magical books, and its contents include instructions for many amulets and magical procedures.

Anonymous
Mirror-image amulet
Cairo Geniza, provenance unknown
Annenberg Research Institute, Philadelphia (Geniza Fragment no. 453)

Protective amulets were prepared for people and virtually everything they owned. Houses, in particular, were often protected by them; in the minds of many Jews, the *mezuzah* affixed to the doorpost of a dwelling similarly ensured the well-being of its inhabitants. Standard rabbinic *mezuzot* contained the text of the *Shema^c* (Deuteronomy 6:4–9, 11:13–21), but many medieval works criticize mystical and magical additions, including unusual scripts and symbols, samples of which were discovered in the Cairo Geniza.

One unusual text is a three-line Hebrew mirror image. Mirrors were a favourite magical tool. Depending on the context, they were used for protection, to reflect various forces, or for divination. Texts written in mirror image sometimes served similar purposes. The first line contains the first verse of the *Shema^c* (Deut. 6:4); the third the rabbinic addition, "Blessed be His glorious name forever." The middle line begins with the words "Holy to the Lord, God of Israel," to which is added a note that the amulet is for an entrance. The text appears to be a dedicatory or prophylactic amulet for an entranceway, the precise use of which may depend on the deciphering of the final word in the second line.

Anonymous
Impotence charm
Italian cursive, seventeenth century
Wellcome Institute Library, London (Heb. MS no. 25)

The mystique associated with one's first sexual encounter was enhanced by the Jewish emphasis on virginity, the extensive attention given to it in rabbinic literature, and the midrashic assertions that several important biblical characters were born of their fathers' first seminal emissions. This charm provides a secret related to male sexual problems.

Mirror-image amulet written to protect an entrance, found in the Cairo Geniza. Annenberg Research Institute, Philadelphia, Geniza Fragment no. 453 (Cat. no. 126)

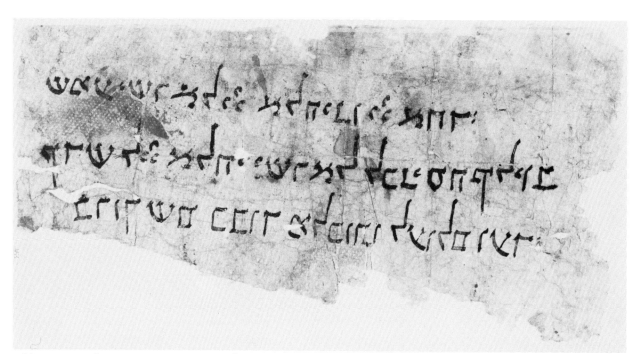

Yiddish and Hebrew prescriptions. Prominently displayed is the word *Abracatabra* (!), in a decreasing pattern which, if uttered properly, caused the illness to disappear. The Wellcome Institute Library, London, MS Heb. A 19 (Cat. no. 118)

Anonymous
Fragment of a medicinal(?) recipe book
Cairo Geniza, provenance unknown
Annenberg Research Institute, Philadelphia (Geniza Fragment no. 457)

Dreams were often believed to convey divine messages; belief in the ability to divine the future through dreams was widespread. The upper section of this fragment contains instructions for consulting a dream. Building on some of the symbolism found in the talmudic lists of dream interpretations, the text suggests that the appearance of sages, synagogues, rabbinic schools, or Jews praying will indicate a positive answer to the question posed before going to sleep, while a negative answer will be shown through dreaming of idolatrous worshippers and temples, unholy places, bath houses, and the like. The text also recommends washing before retiring and wearing pure clothes.

Another concern is the wish one's love be reciprocated. The second part of this text, which suggests that it should be written on a day-old egg and buried under the threshold, contains an Aramaic text. The language resembles that of known Aramaic amulets, but its tone differs from, for example, that of the magic bowls, because it adjures an angel to stimulate love, not to offer protection.

Anonymous
Prophylactic amulet
Iraq: eighteenth or nineteenth century
Ofra Aslan, Montreal

Anonymous
Prophylactic amulet
Iraq: eighteenth or nineteenth century
Ofra Aslan, Montreal

Oriental Jewish communities preserved the practice of wearing large amulets long after the practice was abandoned by most Western groups. Often encased in metal tubes and worn as jewelry, these amulets are frequently very lengthy and contain mystical permutations and combinations of divine names.

Jacob ben Mordecai of Fulda (seventeenth century)

שושנת יעקב
[Shoshanat Ya'aqov]
Amsterdam: 1706
Yeshiva University, New York

The search for meaning in nature and its correlation with human existence led to the interpretation of extraneous or haphazard natural occurences as portents of human situations. In different times and places, this produced among the Jews a belief in the ability to prognosticate from a verse chosen at random from the Bible or recited by a child as his daily lesson, and from the flight of birds, the positions of the stars, and the shapes formed by oil or molten lead dropped into water. Chairomancy, the art of palm reading, often integrated with astrology, was perceived by some to be an adjunct of medical practice. It is discussed in several prominent passages in the *Zohar together with metoposcopy, the related practice of reading the lines on the forehead, which was developed in a number of medieval Hebrew works.

Jacob of Fulda, author of *Shoshanat Ya'aqov*, also wrote *Tiqqun Shalosh Mishmarot*, which was translated and published with his wife's introduction in 1692. *Shoshanat Ya'aqov* claims to derive from the work of a number of scholars, among them Aristotle, who, it says, converted to Judaism.

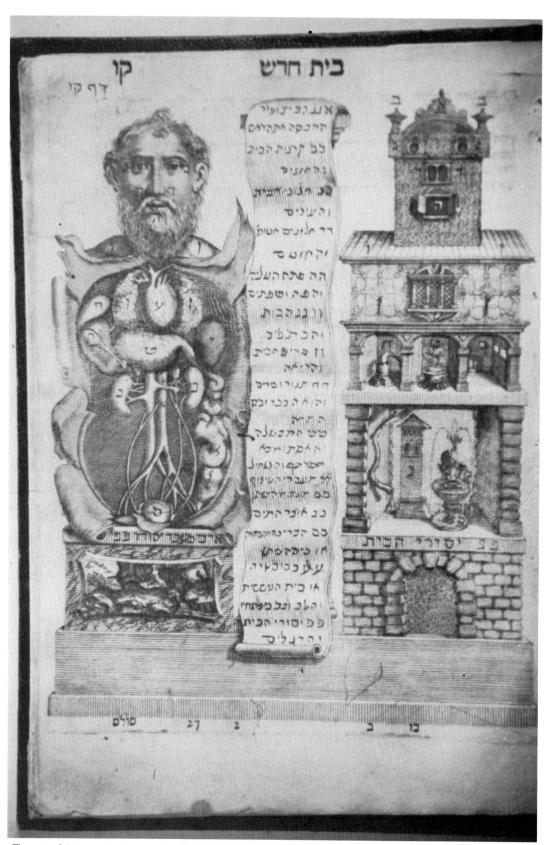

The parts of the human body compared to the parts of a house. Tobias Cohn, *Ma'aseh Tuviah*, Venice, 1708. National Library of Canada, Jacob M. Lowy Collection (Cat. no. 73)

Solomon Almoli

פתרון חלמות ומפשר חלמין
[Pitron Halomot uMefasher Helmin]
Cracow: 1576
Yeshiva University, New York

Dreams are a universally important human experience, and the ancient literature describing and providing directives for interpreting them is quite extensive. Egyptian and Akkadian texts complement Greek works that discuss dream symbols, and the Bible contains enough narratives in which dreams are taken seriously (together with the contemporary non-biblical and Jewish post–biblical parallels) to confirm that the lack of instructions for interpreting them in biblical Israel is not reason to assume that they were ignored there.

Aside from the biblical narratives about Joseph and Daniel, the most important ancient Jewish texts about dreams are found in the Babylonian Talmud. There, we find lengthy discussions of the importance of the interpreter and stories about insincere dream interpreters and their clients. Also contained in these pages are lists of plants, animals, biblical books, ancient kings, and many other things associated with their symbolic values for dream interpretation, as well as a procedure by which a dreamer can "improve" a bad dream by reciting certain prayers and selections from Psalms in the presence of three people who love him.

Despite the fact that several biblical passages, some of the talmudic discussions, and medieval philosophical teachings were quite skeptical about the value of dreams for predicting the future, a rich literature of dream interpretation developed. It shares much with some of the philological elements of midrashic Bible interpretation, and, though not based on studying the stars or the lines of the hands, offers a phenomenon parallel to astrology and chairomancy; modern practitioners often see it as a part of psychology. Regardless of its precise classification, in medieval and early modern times it was treated as a science by many people.

This work, which is highly derivative in nature, includes discussions of the significance of dreams, the ways in which to interpret them, procedures for improving them if they seem to suggest undesirable outcomes, and directions for holding a fast to avert calamity.

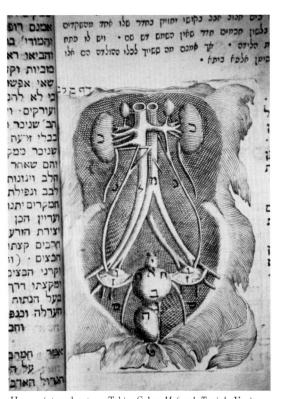

Human internal organs. Tobias Cohn, *Ma'aseh Tuviah*, Venice, 1708. National Library of Canada, Jacob M. Lowy Collection (Cat. no. 73)

Title page from Jonathan ben Joseph of Ruzhany, *Yeshuʿah BeYisraʾel*, Frankfort on Main, 1720. National Library of Canada, Jacob M. Lowy Collection (Cat. no. 136)

Chapter Six

SCIENCE AND RELIGIOUS RITUAL

ונתתי מטר ארצכם בעתו יורה ומלקוש
ואספת דגנך ותירשך ויצהרך:
ונתתי עשב בשדך לבהמתך
ואכלת ושבעת:

דברים י״א, י״ד-ט״ו

And God said, "Let there be lights in the firmament of the sky to differentiate between day and night and to be signs for fixed times and days and years."

Genesis 1:14

One who sees the sun at its turning point, the moon at its strength, the planets in their orbits and the constellations in their order recites: Blessed [are You, O Lord, our God, King of the universe] who does acts of creation.
   And when is this?
   Said Abbaye, "Every twenty-eight years, when the cycle resumes and the Spring Equinox falls in Saturn, between Tuesday evening and Wednesday morning."

Babylonian Talmud, Berakhot 59b

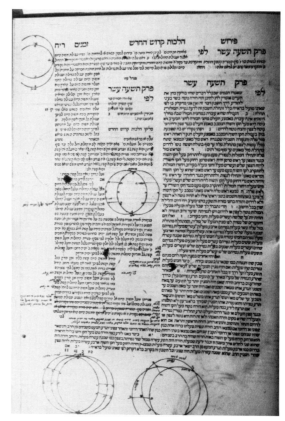

Astronomical sketches from Maimonides' *Hilkhot Qiddush HaHodesh*, Venice, 1550. National Library of Canada, Jacob M. Lowy Collection (Cat. no. 133)

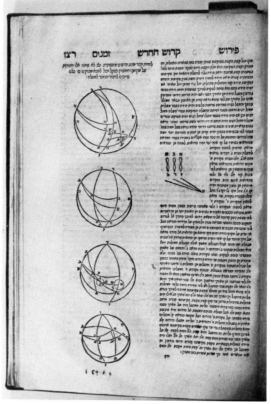

Astronomical sketches from Maimonides' *Hilkhot Qiddush HaHodesh*, Venice, 1574–75. National Library of Canada, Jacob M. Lowy Collection (Cat. no. 134)

THE CALENDAR

Most medieval works of science included full discussions of related religious issues; in the case of astronomy books, the fixing of the religious calendar was an important topic. The history of the calendar is fascinating and complex, and leads through the mythological and scientific literatures of Mesopotamia, Egypt, Greece, and Rome, as well as through medieval and modern attempts to understand and apply them. Having evolved in tandem with many of these other calendars, the Hebrew calendar is almost as complex and no less engaging.

The commonly used names of the biblical months (Tishrei, Adar, etc.) appear only in exilic and post-exilic biblical books because, as the Palestinian Talmud states, "The names of the months came from Babylonia" (*Rosh HaShannah* 1:2). Indeed, all of the names of the months are derived from Akkadian, where they are called *Teshritu, Adaru, Elulu,* and so on. This etymological fact proves the correct name of the month following Tishrei to be Marheshvan, not Heshvan, because the Akkadian equivalent means "eighth month," cognate to *yerah shemini* (marh = yerah). Of similar origin is October, which also means "the eighth month" and falls at the same time of the year; but, because of later additions to the calendar, it is now our tenth month.

The Hebrew calendar counts days according to the solar year, of approximately 365 days' duration, and months according to the lunar one, of approximately 354 days. This causes an annual surplus of approximately eleven days. As this surplus accumulates, it has the potential to push months out of their seasonal cycle (as occurs in the Moslem calendar) but when the total reaches the number of days needed for a full month (after two or three years), a second month of Adar is added at the end of the year, and the remaining days are saved to begin the next surplus. These extra months appear seven times in a nineteen-year cycle (years 3, 6, 8, 11, 14, 17, and 19).

In ancient times, the determination of the months was made visually, and early rabbinic literature is replete with discussions of the proper way to interrogate witnesses who come to testify to their sighting of the new moon, as well as what to do if none appears. It also discusses how to proceed in cloudy weather, how to adjust for sightings that appear at different times of day, when to have one or two days of Rosh Hodesh, the use of travelling emissaries and mountaintop signal fires to notify the entire Middle East that a new moon has been declared, and the manner of determining the holidays by counting from officially announced new moons.

The calendrical system based on lunar observations was eventually discontinued and replaced with one based on calculations. This stimulated the production of a calendrical literature that enabled people in different places to determine if Rosh Hodesh would be one or two days, when it would fall, when leap years would be added, and many other astronomical details related to religious observances. Books of this genre often carry the title *Sefer Ibronot* or the like, and became particularly popular after the calendar reform of Pope *Gregory XIII*.

Moses ben Maimon of Spain and Egypt (1135–1204)

משנה תורה, זמנים, הלכות קדוש החדש

[*Mishneh Torah: Sefer Zemanim, Hilkhot Qiddush HaHodesh*]
Venice: Marco Antonio Giustiniani, 1550
National Library of Canada, Ottawa (Jacob M. Lowy Collection)

133

Venice: Meir Parenzo for A. Bragadini, 1574–75
National Library of Canada, Ottawa (Jacob M. Lowy Collection)

134

Maimonides' complete codification of rabbinic law (*Mishneh Torah*) contains virtually all civil and ritual laws discussed in the antecedent Jewish religious literature and includes detailed plans for such divergent matters as Sabbath and holiday regulations, agricultural laws that applied in Israel, purity laws and sacrifices applicable only in the Jerusalem Temple, marriage and divorce, establishing and conducting the business of the rabbinic courts, and even guidelines for reconstituting a divinely ordained Jewish state. That his code is still one of the pillars on which all rabbinic legal decisions rest is testimony to its universal appeal and status.

Maimonides' work exhibits an holistic approach to knowledge. Law and theology, philosophy and science, medicine and ritual all complement each other in magnificent harmony. It is therefore no surprise that his religious works contain sophisticated scientific notions and that some of his scientific teachings have been invested with religious authority.

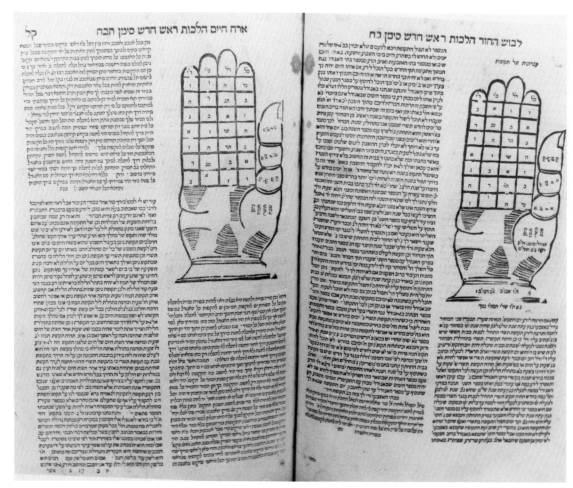

Anonymous anecdote about Maimonides as astronomer. Montefiore Endowment, Jews College, London, MS no. 484 (Cat. no. 135)

Hands used to present calendrical information. Mordecai ben Abraham Jaffe, *Levush HaHor,* Venice, 1620. Jewish Public Library, Montreal (Cat. no. 139)

Hilkhot Qiddush HaHodesh

In ancient times, the beginning of a new month and the dating of all dependent holidays were determined by visual observation of the new moon. But by the early rabbinic period (if not earlier), knowledge of astronomy enabled the sages to predict these events with great accuracy. Even so, there was much room for discussion, and formal astronomy played an important role in Maimonides' codification. One of the important loci of his astronomical discussions is *Hilkhot Qiddush HaHodesh*, "The Laws of the Sanctification of the New Month," which discusses the calendar and, most important, prediction of the visibility of the new moon. It also contains lengthy discussions of the astronomical principles used.

Clarification of Maimonides' astronomical teachings was the responsibility of the many commentators whose interpretations adorned his work. Some undertook this task as a part of the exposition of the *Mishneh Torah* at large; others were specialists who dealt only with *Hilkhot Qiddush HaHodesh* and (possibly) other astronomic passages in his writings.

The commentary accompanying the 1550 folio edition of the *Mishneh Torah* was printed with spaces for the astronomical illustrations, which were added manually, as were minor corrections and additions. Because the spaces sometimes proved inadequate, the margins were used as well, with the result that occasionally the designated spaces have been left blank.

In 1574, the illustrations accompanying the text were printed mechanically for the first time. Later omission of the drawings may have resulted, in part, from the discrediting of the Ptolemaic astronomical system that underlay Maimonides' teachings, an inevitable outcome of the astronomical discoveries of the sixteenth and seventeenth centuries. As well, printers may have avoided illustrations that some readers would have perceived as frivolous in holy books. The few illustrations found in many late editions of *Hilkhot Qiddush HaHodesh* are quite small and are often collected in a brief appendix at the end of the volume.

The 1550 edition of the *Mishneh Torah* is notorious and was a source of widespread antagonism that began as a dispute between two Christian printers in Venice, Marco Antonio Giustiniani and Alvise Bragadini. Rabbi Meir Katzenellenbogen of Padua had hoped to convince Giustiniani, one of the few established Hebrew printers in the world, to issue an edition of the *Mishneh Torah* with his notes. This did not work out, but, he convinced Bragadini, a newcomer to the business, to undertake the effort. Immediately thereafter, Giustiniani published a competing edition of the *Mishneh Torah*, which was met with outrage by Katzenellenbogen and Bragadini. Katzenellenbogen wrote to Rabbi Moses *Isserles, who, together with two other rabbis, placed a ban on Giustiniani's edition *(She'eilot uTeshuvot HaRaMa'* no. 10). Giustiniani responded by seeking help from the Pope. The incident escalated and gave added impetus to the already growing Christian hostility to Jewish religious writings that led to the burning of the Talmud in 1554.

Anonymous
Anecdote about Maimonides
Spanish Rabbinic Script, sixteenth–seventeenth century
Montefiore Endowment, Jews College, London (MS no. 484)

135

Montefiore manuscript no. 484 contains a story about Maimonides' astronomical activities. This anecdote, attributed to one of his students, reports that, while engaged in his astronomical pursuits, the famed rabbi received a call to appear before a certain court official (*sar*). Maimonides declined, saying that he was overcome by his own insignificance, highlighted by his relative lack of size and importance compared with those of the universe he was studying. Under the circumstances, he continued (perhaps only half-seriously), there would be no sense to his appearing before the official, whose glory so surpassed his own that he probably would not realize he had come.

Jonathan ben Joseph of Ruzhany (seventeenth–eighteenth century)

ישועה בישראל
[Yeshuʿah BeYisra'el]
Frankfort on Main: Johann Koelner, 1720
National Library of Canada, Ottawa (Jacob M. Lowy Collection)

136

Jonathan ben Joseph of Ruzhany was a talmudist and astronomer who wrote commentaries on important medieval works of science, including *Maimonides' *Hilkhot Qiddush HaHodesh*, Abraham *Bar Hiyya's *Tzurat HaAretz*, and the Hebrew translation of *Sacrobosco's *Sphaera mundi*.

His work on Maimonides discusses the astronomical portions of the *Mishneh Torah*, including the third chapter of *Sefer HaMadaᶜ* and all of *Hilkhot Qiddush HaHodesh*. Exhibiting the influence of *Delmedigo, *Yeshuʿah BeYisra'el* recognizes the contributions of Galileo and *Copernicus. It includes a full commentary on Maimonides, accompanied by many large and impressive illustrations. In one case (p. 44b), a second drawing has been pasted over the printed one. The volume also contains the glosses of Samson ben Bezalel (brother of the Maharal of Prague) on Abraham bar Hiyya's *Tzurat HaAretz*.

Raphael ben Jacob Joseph HaLevi of Hanover (1685–1779)

תכונת השמים
[Tekhunat HaShamayim]
Rabbinic cursive, eighteenth-nineteenth century
Annenberg Research Institute, Philadelphia

137

Amsterdam: Jahn Janson, 1756
Jewish Public Library, Montreal

138

Raphael was born in Weikersheim, Franconia. He was employed as a bookkeeper, but spent a number of years studying with Leibnitz, who recognized him as an outstanding student. He also taught mathematics and astronomy and composed a number of astronomical works. The best known of these is *Tekhunat HaShamayim*, which deals with fixing the calendar and other astronomic matters of halakhic import and is augmented by extensive tables. Among its goals is clarification of *Maimonides' *Hilkhot Qiddush HaHodesh*.

Since it is based on Maimonides, *Tekhunat HaShamayim* follows the geocentric, Ptolemaic system; but, at the end of the work, Raphael discusses and acknowledges the validity of the Copernican thesis, for which an illustration is provided. The title page indicates that the book has been augmented by Moses Tiktin, his student, but, according to A. Neher, *Tekhunat HaShamayim* comprised Raphael's lecture notes, edited by Tiktin without his teacher's approval.

Raphael's other works include *Hokhmat HaTekhunah*, on astronomy; *Luhot HaIbbur* (Leyden–Hannover: 1756), consisting of extensive astronomical tables; and *Kelalei Sod HaIbbur*.

Mordecai ben Abraham Jaffe of Prague, Venice, Grodno, Lublin, Kremeniec, and Posen (ca. 1535–1612)

לבוש החור
[Levush HaHor]
Venice: Petrus and Lorenzo Bragadini, 1620
Jewish Public Library, Montreal

139

לבוש אדר היקר
[Levush Eder HaYaqar]
Lublin: Kalonymos ben Mordecai Jaffe, 1595
Yeshiva University, New York

140

Mordecai *Jaffe studied in Poland under Solomon Luria and Moses *Isserles. His education included Kabbalah as well as philosophy and classical rabbinic texts, and the influence of his rabbinic contemporaries, including Isserles, is evident in his work on astronomy. His many compositions reflect a wide range of interests and include commentaries on Rashi's midrashic Pentateuch commentary, Recanati's kabbalistic Pentateuch commentary, *Maimonides' *Guide for the Perplexed*, and Abraham *Bar Hiyya's *Tzurat HaAretz*. Jaffe's books, all containing the word *levush* ("garment") in the title, are known collectively as the *Levushim* and consist of ten titles in eight volumes.

The most famous *Levushim* are the five that describe the areas of religious practice covered in Jacob ben Asher's *Arbaʿah Turim and Joseph Caro's *Shulhan Arukh. Stylistically, they fall between the two preceding works, avoiding both the lengthy discussions of the former and the terse codification of the latter. Substantively, they add legal material omitted by both. *Levush HaHor,* one of these, is the part of the code that deals with Sabbaths and holidays, including *Rosh Hodesh* and the fixing of the calendar. Another, *Levush Eder HaYaqar,* is similar in nature to Maimonides's **Hilkhot Qiddush HaHodesh* and includes illustrated astronomical discussions as well as comments on Abraham Bar Hiyya's *Turat HaAretz.

David ben Joseph Abudarham of Seville (fourteenth century)

פירוש הברכות והתפילות
[Peirush HaBerakhot VeHaTefillot]
Lisbon: Eliezer Toledano, 15 November, 1489
National Library of Canada, Ottawa (Jacob M. Lowy Collection)

141

David ben Joseph Abudarham is reputed to have been a student of *Jacob ben Asher, because he frequently mentions teachings associated with Rabbeinu Asher and his son Jacob. Whether or not this is so, he was, in any event, an important and respected community leader in Toledo, where they lived.

Peirush HaBerakhot VeHaTefillot contains detailed discussions on many aspects of the Hebrew liturgy and related halakhic practices, including the construction of the altar in the Temple; recitation of the *Shemaʿ* and *Amidah,* the daily, Sabbath, and holiday prayer services; the Passover Haggagah; passages from the Mishnah that appear in the liturgy; and the thirteen hermeneutical rules of Rabbi Ishmael. Abudarham's scientific interests are evident in his explanation of rules of intercalation and in his calendrical and astronomical tables, which do not appear in all editions of the work.

Abudarham's work is an important repository of early responsa of the Babylonian *geonim*. His impact on the history of liturgical interpretation and practice was significant, as evidenced by the extent to which his teachings are cited in later halakhic works and by the number of times the book was printed.

This copy of *Peirush HaBerakhot VeHaTefillot,* one of the few remaining samples of the second Hebrew book printed in Lisbon, may be the only complete copy in existence.

ספר עברונות
[Sefer Ibronot]
Rabbinic Script, late seventeenth century
Hebrew Union College, Cincinnati (MS no. 906)

142

Ashkenazi cursive, ca. early nineteenth century
Hebrew Union College, Cincinnati (MS no. 903)

143

Lublin: Zvi Jaffe, 1640
National Library of Canada, Ottawa (Jacob M. Lowy Collection)

144

Rabbinic hand, late seventeenth century
Columbia University, New York (X893 Se36)

145

Eliezer ben Jacob Ashkenazi Belin (sixteenth-seventeenth century)
Frankfort on the Oder: Johann Christoph Beckmann, 1601
Annenberg Research Institute, Philadelphia

146

Meir ben Nathan Joshua Himmelburg
Offenbach: Israel ben Moses, 1722
Yeshiva University, New York

147

Many books with the title *Sefer Ibronot* (or something very similar) have been published over the centuries, and books in this genre are among the most popular and inexpensive types of Hebrew manuscripts held in many public and private collections. Typically, a *Sefer Ibronot* contains rules for fixing the calendar, charts, and some halakhic data of calendrical import. More embellished copies have decorations, and the charts are coloured or organized in more attractive, and sometimes less functional, ways. The most ornate copies are extensively illustrated in colour. Human hands and hand-held balances or

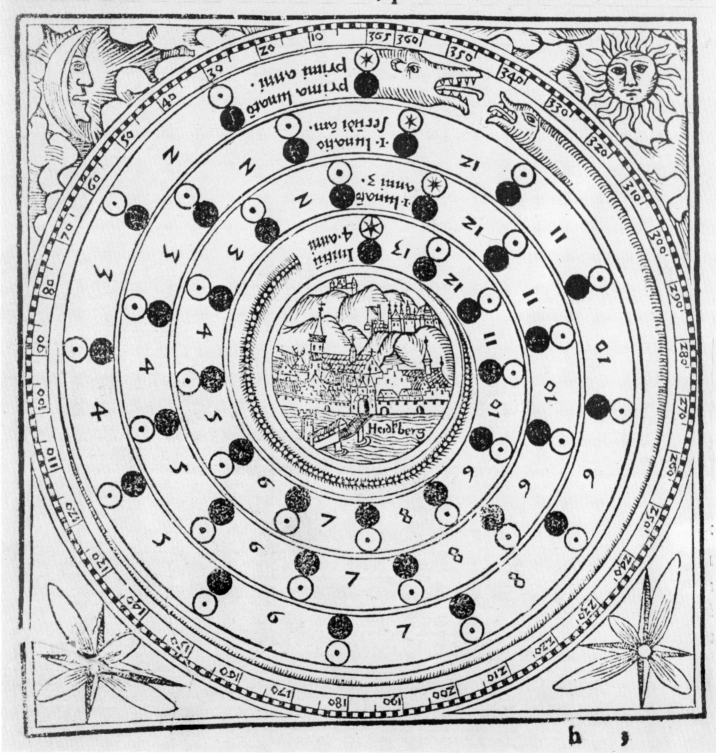

A spiral correlation of the days and months of four lunar and solar years. Note that the fourth and innermost year contains thirteen months, which helps align them. Sebastian Muenster, *Kalendarium Hebraicum, Opera Sebastiani Munsteri ex Hebraeorum penetralibus...*, Basle, 1527. Yeshiva University, New York (Cat. no. 139)

scales are two dominant artistic motifs.

Hebrew Union College manuscript no. 906 was copied and illuminated by Zechariah Shimon ben Jacob from a Warburg family. In addition to the customary calendrical materials, it contains sections with the formal texts of legal documents, medical recipes, and suggestions for influencing the weather and making a cock crow.

The first edition of Jacob Marcaria's *Sefer Ibronot* appeared in 1560 and was followed by a series of editions that adopted, adapted, and augmented the material. The 1691 edition, compiled by Belin, was printed by Beckmann, a Protestant historian and philologist who owned the first Hebrew press in Prussia. The Columbia University manuscript of *Sefer Ibronot* contains a number of interesting illustrations, in addition to the standard calendrical materials and related miscellanies. Adam and Eve appear clothed in fig leaves, and, together with the execution of Agag and scenes from the lives of Jepthah and Isaac, symbolize the four seasons.

David ben Raphael Meldola (eighteenth century)

מועד דוד

[Moʿed David]
Amsterdam: Abraham ben Raphael Hezkiah Athias, 1740
Yeshiva University, New York

| 148 |

Moʿed David is similar to many of the *sifrei ibronot*, but it is slightly more focused on astronomy.

Sebastian Muenster of Basle (1489–1552)
Kalendarium Hebraicum, Opera Sebastiani Munsteri ex Hebraeorum penetralibus ...

חכמת המזלות בתקופות ומעוברות והקביעות

[Hokhmat HaMazalot BaTequfot uMeʿubarot VeHaQeviʿot]
Basle: Hieronymus Froben, 1527
Yeshiva University, New York

| 149 |

This volume contains a collection of materials edited and translated into Latin by Sebastian *Muenster. Included are astronomical and calendrical sections, such as "*De anno et mensibus Hebraeorum,*" "*Tractatus de inventione neomeniarum et aequinoctiorum,*" and "*Tabula excessuum,*" which treat the dynamics of fixing the lengths of the months and related lunar variables that determine the exact nature of the calendar. Also included are parts of *Seder Olam Zuta* and Abraham ben David's *Sefer HaQabbalah*, an important medieval history of the Jews. The book is very attractively illustrated.

CIRCUMCISION

According to the Bible, circumcision has been a hallowed practice since the time of Abraham. Over the centuries, this surgical procedure, performed on boys at the age of eight days, has developed into Judaism's most essential male rite of initiation. Though it was practised by other ancient peoples, including the Egyptians – from whom, it has been suggested, Abraham learned of the procedure – circumcision became the permanent sign of the covenant between Jews and God. It has been practised by Jews continually since Antiquity, and literary sources from the Greco-Roman period testify to the tremendous sacrifices many Jews made to observe it, often including the cost of their own lives.

Since biblical times, when flint knives were used, a variety of medical improvements have found their way into the circumcision procedures, and the ceremony itself has continually evolved. Indeed, a full history of the surgical equipment and accompanying prayers is recoverable. Halakhic discussions of circumcision also contain interesting notes on jaundice (which, even if relatively mild, would delay the ceremony), agents used to help the blood coagulate, and hemophilia.

Except in unusual circumstances, the medical procedures for circumcision are quite simple, as are the accompanying rituals. Yet, despite the ancient tendency for fathers to circumcise their own sons, many communities designated specialists for this purpose, especially in light of the belief that spiritual benefits derived from having it performed by a particularly pious individual.

Mohalim, ritual circumcisers, often used small handbooks containing extracts of the

Discussion of circumcision, Z. Grapius, *Controversiam recentissimam historicam an circumcision ab Aegyptiis*. The Wellcome Institute Library, London (Cat. no. 152)

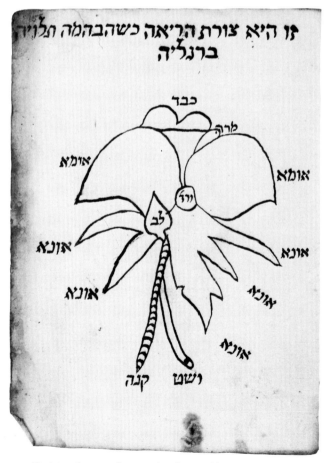

The internal organs of an animal as they would appear in a carcass suspended upside down. Jacob Weil, *Derekh HaBediqah*, The Wellcome Institute Library, London, MS Heb. A 7 (Cat no. 153)

relevant laws, the prayers recited during the ceremony, and the accompanying procedures for naming the child. They were also frequently used to record the names of the boys who had been circumcised. Occasionally, the books were decorated with colour illustrations.

David ben Aryeh Leib of Lida (ca. 1650–96)

סוד ה׳

[Sod Adonai]
Amsterdam(?): Dov Katz, 1687
Square, rabbinic script with vocalization
Wellcome Institute Library, London (MS no. 4)

150

David ben Aryeh was born in Lida, and served as rabbi in a number of European communities, including Mainz and Amsterdam. He contributed to the Jewish responses to Eisenmenger's anti-Semitic attacks, but he was not well liked, even by Jews. In Amsterdam he was criticized for taking *Migdal David,* his commentary on Ruth, from Hayyim ben Abraham HaKohen's *Torat Hesed* and for being a Sabbatian sympathizer.

In both cases he was exonerated, but these criticisms were not without some justification, and the accusations left their mark. David was forced to change rabbinic positions on a number of occasions, and the title of a posthumous collection of his works, *Yad Kol Bo,* reflects his oppressed feeling. The phrase, traditionally understood to be negative, was applied to Ishmael by an angel in Gen. 16:12. Here *Yad* (which also represents the Hebrew number fourteen) refers to the fourteen sections of David ben Aryeh's book, and *Kol Bo* to its anthological nature. But the hostility and frustration felt by the author are also apparent from the quotation, which means "everyone's hand is against him."

Sod Adonai was published many times, often accompanied by a commentary entitled *Sharvit HaZahav,* and became one of the standard manuals used by *mohalim*. The Wellcome Institute manuscript of eighteen pages contains twelve very attractive miniatures that depict various stages of the liturgical and medical aspects of the circumcision.

ספר אות הברית

[Sefer Ot HaBerit]
Rabbinic hand, 1824
Columbia University, New York (X893 T71)

151

This small attractive circumcision handbook was a gift presented to Yehi'el Poresh, a physician and ritual circumciser in Leipnik. It contains an interesting micrographic representation of King Solomon, who symbolized the wise ruler and master of scientific knowledge.

Zacharias Grapius (1671–1713)

Controversiam Recentissimam Historicam an Circumcisio ab Aegyptiis ad Abrahamum Fuerit Derivata? a Jo. Marshamo Anglo Potissium Jo. Spencero Motam examinat, Eamque Benevolo Superiorum Consensu In Illustri Academia Rostochiensi Praeside Viro Plurimum Reverendo Excellentissimo et Amplissimo Domino Zacharia Grapio
Jena: Io. Bernard Hellerum, 1722
Wellcome Institute Library, London

152

Though it found many aspects of Judaism appealing, the ancient Hellenistic world balked at circumcision. Scholars have often suggested that the adoption of Christianity rather than Judaism was facilitated by the former's abolition of this unacceptable requirement. More modern debates have centred on claims that circumcision aids in preventing disease, including cervical cancer, and its alleged ability to both strengthen sexual self-control and heighten coital pleasure.

The origins of circumcision have also come under scrutiny. While it is popularly perceived as a Jewish practice, it is also observed among other Semitic groups and was practised by the ancient Egyptians, as evidenced by the discovery of statues of circumcised males. Abraham was the first biblical character noted to have practised circumcision, and it is clear that he did so only after his trip to Egypt, which suggests the possibility of an Egyptian origin.

It is impossible to demonstrate conclusively that circumcision was borrowed from the Egyptians, but many Israelite practices do relate to those of the surrounding nations. Agricultural celebrations, the use of animals for sacrifices, worship of God in specific types of places, the stylistic rules for biblical poetry, and much of biblical law are too similar to ancient parallels to have developed in total isolation. Rather, these and the many similar cases, including circumcision, must be understood as examples of adopting and adapting ancient institutions for new religious purposes. Circumcision was presented in the Bible as the sign of the covenant between God and Abraham and his descendants, and this remains the basis for appreciating its ancient significance. Neither foreign parallels nor any potential psychological or medical benefits are relevant to its biblical presentation.

One of the many formal discussions of the origins of circumcision held over the centuries was conducted under the supervision of Z. Grapius, a Lutheran Hebraist. Note the rabbinic sources cited in the Latin presentation.

RITUAL SLAUGHTER OF ANIMALS FOR FOOD

Jewish dietary laws are extensive and include prohibitions against eating certain foods, especially certain types of fish, birds, and animals (and their derivatives), and against mixing some permitted foods, for example, milk and meat (and their derivatives). The dietary laws also impose detailed requirements on the method of slaughter of permitted animals, inspection of their internal organs, and removal of blood from their flesh before consumption through a process of soaking and salting. These laws often combine both physics and metaphysics, but the prominence of rules and decisions based on pure scientific data (biology, chemistry, or physics) and mathematical analysis (the application of the laws of probability) means that the study of these ritual requirements and their related problems – and the halakhic literature that deals with them – is largely scientific in nature.

The rules of kosher slaughter, which derive in part from the ancient procedures of animal sacrifice, require among other things use of a very sharp knife, extreme care to avoid pulling or tearing the incision, and inspection of the internal organs to certify the health and fitness of the animal. They are discussed at length in the Mishnah and Talmud, and are essential concerns of all of the major halakhic codes –*Mishneh Torah, *Arba῾ah Turim, *Shulhan Arukh, and *Levush, to mention a few. They have also been presented in a series of specialized works devoted only to this topic.

Jacob [ben Judah?] Weil of Nuremberg, Augsburg and Erfurt (fourteenth-fifteenth century)

הלכות שחיטה ובדיקות
[Hilkhot Shehitah uBediqot]
Ferrara: Italian cursive, 1718
Wellcome Institute Library, London (MS no. 7)

153

בדיקות
[Bediqot]
Mantua: Jacob Cohen, 1560
McGill University, Montreal

154

Jacob Weil was an important medieval German rabbinic figure whose literary contribution consists of one volume of responsa to practical halakhic questions and several very small and often appended halakhic compositions. A rigourous legal thinker, Weil rarely cited his well-known Sefardic predecessors, preferring, instead, the legal precedents and writings of Ashkenazim.

Hilkhot Shehitah uBediqot is a very brief two-part composition written for ritual slaughterers. It contains the minimal instructions for proper slaughter and inspection of the animal and assumes some prior acquaintance with the concepts and their applications. The first part, which deals with the slaughtering process itself, is devoted primarily to the five human errors that can invalidate the procedure. The second part outlines the procedures for inspecting the carcass and determining the presence of specified anatomical abnormalities that would prohibit its consumption. Occasionally, the latter work is accompanied by drawings of the entrails that clarify the descriptions.

Hilkhot Shehitah uBediqot was published many times as a booklet, sometimes in two separate parts. It was often accompanied by commentaries composed by very prestigious rabbis, including Moses *Isserles and Solomon Luria.

Anonymous

הלכות שחיטה
[Hilkhot Shehita]
Square script, eighteenth-nineteenth century
Wellcome Institute Library, London (Heb. MS no. 8)

Yiddish has been in use for about one thousand years, but its period of greatest strength was before World War II, when approximately eleven million people spoke it. It was an important medium both for the teaching of religion and for secular learning, including folk medicine. While most rabbinic works were composed in Hebrew, Yiddish usually served as the language of halakhic instruction. Brief digests of specific topics in religious law were often composed in Yiddish, of which Wellcome manuscript no. 8, on the kosher slaughtering of animals, is an example.

Elhanan David Carmi

הלכות שחיטה ובדיקה
[Hilkhot Shehita uBediqah]
Eliezer ben Mordecai, Italian rabbinic cursive, 1715
Columbia University, New York (X893 CZ1)

The practice of composing small handbooks like that of Rabbi Jacob Weil for those who engage in ritual slaughter was so widespread that some bibliographers refused to list the various texts known to them. One such item is the text of Elhanan Carmi, which includes a routine drawing of an animal's lung.

Anonymous
Karaite work on ritual slaughter
Eighteenth or nineteenth century
Columbia University, New York (X893 H542)

One of the primary tenets of Karaite doctrine was the rejection of the Oral Torah, the corpus of teachings that the rabbis taught as the correct and binding interpretation and application of the Torah given at Sinai and that the Karaites perceived as late rabbinic fraud. Initially, Karaite leaders set out to base their religious teachings on meticulous and far-reaching exegesis of the Bible, but their efforts proved extremely difficult and ultimately led to the creation of Karaite traditions that parallelled the rabbinic ones.

As a movement, Karaism is very committed to its own type of religious practice, and comparisons of Karaite and rabbanite interpretations of biblical law and practice are both interesting and edifying. This manuscript reflects the Karaite concern for proper ritual slaughter of animals, and contains a series of questions that would be asked of candidates for the position of ritual slaughterer.

Moses ben Gershom Gentili (Hefez) of Venice (1663–1711)

מלאכת מחשבת
[Melekhet Mahshevet]
Venice: 1710
McGill University, Montreal

Gentili was a rabbinic scholar in Venice who, like many of his colleagues, was involved in mathematics, philosophy, and science. He wrote *Hanukkat HaBayit*, on the construction of the Temple, and several other pieces, but his major work, *Melekhet Mahshevet*, is a philosophical-homiletical commentary on the Torah, accompanied by tables and several interesting pictures.

It was quite normal for authors' portraits to be included in books published in the seventeenth and eighteenth centuries, but the one accompanying *Melekhet Mahshevet* is unusual for several reasons. The caption seems to indicate that, at the time, the author was one hundred years old, but in 1710 Gentili was under fifty. In fact, the numerical values of the three letters of the Hebrew word *me'ah* add up to forty-six which was his age at that time.

The second edition of *Melekhet Mahshevet*, which appeared in Koenigsberg in 1819, contains a similar portrait and caption (reproduced in the *Encyclopaedia Judaica* 7:414), but here the subject's hair is greyer, his features are more stylized, and he is wearing a *kippah*.

The different angles of vision from the deck and crow's nest of a ship; note the sizes and shapes of the continents, especially Antartica. Tobias Cohn, *Ma'aseh Tuviah*. Venice, 1708. National Library of Canada, Jacob M. Lowy Collection (Cat. no. 73)

Chapter Seven

GEOGRAPHY AND CARTOGRAPHY

> הִתְבּוֹנֵן עַד רַחֲבֵי אָרֶץ
> הַגֵּד אִם יָדַעְתָּ כֻלָּהּ׃
>
> איוב לח, יח
>
> לְדָוִד מִזְמוֹר לַה' הָאָרֶץ וּמְלוֹאָהּ
> תֵּבֵל וְיֹשְׁבֵי בָהּ׃ כִּי הוּא עַל יַמִּים
> יְסָדָהּ וְעַל נְהָרוֹת יְכוֹנְנֶהָ׃
>
> תהלים א, ב

In the book of Rav Hammuna, the elder, it is explained further that the earth (*yishuva'*) is shaped like a sphere, [with] some [inhabitants] below and some above. All the creatures of unusual appearance [i.e., the various races] result from the differences in the atmosphere of each individual place, and their life expectancies are like the rest of mankind [in contrast to those of the creatures mentioned several lines earlier, who live for only ten years].

Moreover, there is a place on the earth where it is dark for some people, while it is light [elsewhere] for others; for some it is day, and for others night. And there is a place that is always daylight and never night, except for a short time. What is reported in the ancient books and the Book of Adam is correct... The secret was transmitted to the masters of wisdom ...

Zohar III, 10a

The passage in the Zohar, Mantua, 1558–60, that reveals a knowledge of living conditions in various places around the globe. National Library of Canada, Jacob M. Lowy Collection (Cat. no. 159)

Men standing at the north and south poles of the earth with their feet pointing toward the centre. Tobias Cohn, Ma'aseh Tuviah, Venice, 1708. National Library of Canada, Jacob M. Lowy Collection (Cat. no. 73)

Moses ben Shem Tov de Leon of Castile (ca. 1240–1305)

ספר הזהר
[Sefer HaZohar]
Mantua: Meir ben Ephraim and Jacob ben Naphtali HaKohen, 1558–60
National Library of Canada, Ottawa (Jacob M. Lowy Collection)

159

Some ancient thinkers believed that the earth was shaped like a sphere, but the popular belief that it was disk-like was more influential, and only in the late fifteenth century did the spherical configuration gain popular support in the West. Even so, a number of early rabbinic sources refer to the spherical form. The Palestinian Talmud, for example, states that "the world [i.e., the earth] is shaped like a sphere [*kadur*]." And a legend reports that Alexander the Great went aloft into the sky until he saw "the world [again, the earth] like a sphere [*kadur*]" (*Avodah Zarah* 42c).

Medieval writers, including some astronomers whose works are presented in chapter 2, discussed the qualities of a spherical earth, and a very interesting reference is found in the Zohar, one of the most influential Jewish books ever composed. Written in Aramaic and attributed to Simeon bar Yohai, a first-century rabbi, but now believed to be the creation of Moses de Leon, who lived in thirteenth-century Spain, the Zohar appeared to be older than, and hence more authoratative than, the *Mishnah and the *Talmud, the quintessential rabbinic guides to religious thought and life. Its spiritual focus and esoteric qualities endeared it to generations of devoted students.

The debate on the Zohar's antiquity raged among both Jewish and Christian scholars several centuries ago, and even some traditionalists recognized that it was not completely authentic. Despite these doubts, its influence on the evolution of Jewish thinking has been profound and pervasive since the sixteenth century.

The Zohar is not a work of science; indeed, many enlightened critics perceived its speculative excesses to be contrary to scientific thinking. But this attitude was far from universally accepted before the dawn of the Enlightenment, and many of its passages must be appreciated in the context of the scientific explorations and speculations of ancient and medieval times. Particularly during the sixteenth and seventeenth centuries, when zoharic influence was at its apex, much effort was devoted to synthesizing its world view with the emerging scientific ones; and those passages that supported the new scientific "truths" were often paraded as evidence of the book's antiquity and divine origins.

In this context, it is interesting to observe that the Zohar refers to the spherical shape of the earth. In a discussion of Genesis 1:6, "Let there be a firmament in the midst of the waters," the Zohar reports that the earth is shaped like a sphere; that some of its inhabitants live above, and others below; that there is a place on the earth where it is dark for some people, while it is light elsewhere for others; that for some people it is day while others have night; and that a place exists where daylight is almost constant.

The Zohar was maintained as an esoteric tradition taught to initiates, and serious debates surrounded the propriety of publishing it. In the end, it appeared almost simultaneously in two editions: Mantua, 1558–60, and Cremona, 1559–60.

Gabriel de Vallsecha (fifteenth century)
"The Mediterranean Sea and the Black Sea"
Majorca: Gabriel de Vallsecha, 1447
Bibliothèque Nationale, Paris (Res. Ge C no. 4607)

160

The best maps produced during the fourteenth and fifteenth centuries were portolans, which are drawn on a pre-planned grid consisting of interlocking squares, rectangles, and triangles. An accompanying handbook of distances, collected from reliable travel records, provided data for navigators and enabled them to calculate very accurately, by means of simple geometric formulae, the distances between spots on the map.

Hayyim ibn Rich (thought by some to have been a relative of Abraham Crescas, a very famous Jewish cartographer) was baptised during the massive forced conversion of 1391. He took the name Juan de Vallsecha, and both he and Gabriel de Vallsecha, assumed to be his son, were famous Majorcan mapmakers.

Gabriel produced a map of the world used by Amerigo Vespucci, as well as a very accurate map of the Mediterranean and Black seas. The latter evidently served primarily for navigation, as the vast majority of labelled cities are on the coasts. In addition to the many names that highlight the shoreline, the map is attractively decorated with colour banners leading to many of the names. It also contains representations of buildings in eight cities, including Cairo, Damascus, Venice, and Grenada. Jerusalem is represented by the holy sepulchre, a standard symbol in maps of this period.

Judah ibn Zara (Abenzara) (fl. 1500)
"Map of the Mediterranean"
Alexandria: Judah ibn Zara, 1497
The Vatican (Borgiano vii)

161

Judah ibn Zara, 1500
Hebrew Union College, Cincinnati

162

Little is known about Judah ibn Zara, also of Majorca, but three of his maps have survived. The one in the Vatican is dated Alexandria, 1497. A second is inscribed "Safed in Galilee, October 1505." The only one in North America, owned by Hebrew Union College, dates from 1500.

Benedict Arias Montanus
Sacrae Geographiae Tabulam ...
Antwerp: 1571
David M. Stewart Museum, Montreal

163

Arias Montanus was an important mapmaker who contributed the geographic materials for a sixteenth-century polyglot Bible in Hebrew, Greek, Latin, and Syriac. His map of the world, which is essentially a graphic interpretation of the tenth chapter of Genesis, locates Noah's descendants around the globe.

Descendants of Japhet are labelled with Roman numerals, those of Shem with Latin letters, and those of Ham with Arabic numbers. Among the more interesting are the placing of Seba, son of Cush (Gen. 10:7) in Siberia, and the placing of Jobab (Gen. 10:29) and Sefar (Gen. 10:30) in South America. Also noteworthy is the double placement of Ophir (Gen. 10:28) on the western coasts of both North and South America.

While the precise location of this Ophir is in doubt, Ophir is also the name of the place from which Solomon's sailors brought large quantities of gold (e.g., *I* Kings 9:28). Using a variety of evidence and arguments, Cyrus Gordon has tried to prove that the Phoenicians sailed to the New World, and others have even suggested a possible connection between the names Peru and Ophir. However accurate or fanciful Gordon's reconstruction eventually proves to be (and thus far it has found little support from contemporary historians), it is clear that sixteenth-century Christian cartographers took seriously the biblical description of the spread of mankind as described in Genesis 10 and assumed that people travelled from the Middle East to South America at least once.

Heinrich Buenting (1545–1606)
Itinerarium Biblicum Sacrae Scripturae
Helmstadt: Jacob Lucius Siebenbuerger, 1581
David M. Stewart Museum, Montreal

164

While many cartographers laboured to make their maps as accurate as possible, others drew them schematically. One popular medieval pattern, known as a "T-O map," divided the world into three parts by depicting it as a T circumscribed with a circle. The space above the horizontal top of the T corresponded to Asia, while the areas to the left and right of the vertical line represented, respectively, Europe and Africa.

Carrying the graphic symbolism even further, Heinrich Buenting, a well-known Protestant minister, sketched Europe as a virgin and Asia as Pegasus, a mythological winged horse. He also popularized the presentation of the inhabited world as a cloverleaf in his biblical travel book. In the latter case, three equally spaced oval leaves corresponding to Asia, Africa, and Europe surround a centre, and the New World appears in the lower left corner. Aside from its obvious artificialty, the most noteworthy aspect of this map is Jerusalem's place as the relatively prominent centre. Jerusalem's location is, indeed, very near the crossroads of the three continents, which contributed to four thousand years of rotating conquest of the city by various European, Asian, and African powers. But the map's theological and political perception of Jerusalem as the centre of the world is unavoidable, and this notion has endeared the map to Jews and ensured its frequent reproduction.

Mordecai ben Abraham Jaffe of Prague (ca. 1535–1612)
לבוּשׁ הָאוֹרָה
[Levush HaOrah]
Prague: Hayyim ben Jacob HaKohen, 1604
National Library of Canada, Ottawa (Jacob M. Lowy Collection)

165

The sixteenth century was a time of extensive explication of Rashi's Torah commentary, and *Levush HaOrah*, by Mordecai *Jaffe, has become one of the most popular super-commentaries of its kind. A number of sixteenth-century commentaries on Rashi are accompanied by maps or sketches, and *Levush HaOrah* is no exception. The Venice, 1527, edition of the super-commentary by Elijah *Mizrahi contains such a map, and a primitive cartographic sketch is also found in the Prague, 1612, edition of Moses Mat's *Ho'il Moshe* (p. 31a). Jaffe's map, is quite similar to both Mizrahi's and the less ornate one in a manuscript of Rashi's commentary (Munich, Staatsbibliothek, MS no. 5, Vol. 1, p. 140a) dated 1233. While it is thus true to its goal of interpreting Rashi, it does not draw upon the more advanced geographic information available during Jaffe's lifetime.

Joseph ben Issachar Ber of Prague (end of sixteenth century)
יוֹסֵף דַּעַת
[Yosef Daʿat]
Prague: Gershom ben Bezalel Katz, 1609
National Library of Canada, Ottawa (Jacob M. Lowy Collection)

166

Yosef Daʿat, is an illustrated commentary on Rashi's Torah commentary that also contains a map. While some of the cities marked are questionably located, the identities of the bodies of water are certain. It is therefore interesting to note the positions of the Jordan River, the Dead Sea, and *Yam Kinneret* (Sea of Galilee). The label *"yam yam"* on the western edge of the map, which refers to the Mediterranean Sea, is *yam hagadol* in other maps of the period, and may reflect a copying error.

Somewhat typical of medieval works (but perhaps unexpected for modern readers) are the location and relative size of the *Yam Suf* (Red Sea). Also of note is the manner in which its crossing has been depicted. Clearly, the artist perceived this as entering into the sea and leaving it on the same bank, rather than crossing from one side to the other. This interpretation parallels that offered by the tosafistic gloss to *Arakhin* 15a (a discussion of Ex. 14:29), in which a similar picture appears in the Romm edition of the Babylonian Talmud. A related notion has been suggested by some modern interpreters, who understand the route to have been along a semi-circular path jutting into the Mediterranean.

Mizrahi's map (and, indirectly, the others related to it) was criticized as early as the sixteenth century by an Egyptian rabbi, David ben Solomon ibn abi Zimra, who objected to equating the biblical *Nahal Mitzrayim* with the Nile River, and correctly identified it as Wad(i) El Arish.

Johann Baptist Homann (1663–1724)
Asia polyglotta linguarum genealogiam...
Nuremberg: 1753
David M. Stewart Museum, Montreal

167

As European discoveries of the earth's land masses and their inhabitants proceeded, cartographers, linguists, ethnographers, and other scholars rapidly assimilated the new findings. A series of four maps of Asia, Europe, Africa, and America designed by Johann Baptist Homann, a German cartographer and publisher, depicts the intersection of these interests. The maps include mountain ranges and large bodies of water, and each geographic area is labelled in Latin characters and marked by a few words in the local language and script. Presentation is further enhanced by alphabet charts that offer transliterations in Latin and, in some cases, the names of the corresponding letters.

The map of Asia covers the entire continent, and contains labels in many different scripts; its linguistic focus is clearly on the Semitic languages. Five of the eight alphabet charts are Semitic – Hebrew-Chaldaean (Aramaic), Samaritan, Syriac, Arabic-Persian, and what has been called Palmyrene – as are all three presentations of vocalization, but there is no guide to the oriental languages depicted on the map (Sanskrit, Chinese, and Japanese), though some space on the accompanying maps is devoted to them. The Hebrew-Aramaic script is placed on Israel, the Arabic-Persian in the Arabian peninsula and east of the Tigris, and the Syriac in western Syria. No place is labelled with either the

Skindiving in Calcutta. Abraham Farissol, *Iggeret Orḥot Olam,* Prague, 1793. Annenberg Research Institute, Philadelphia (Cat. no. 168)

"Sailors who work at sea witnessed God's creatures and wonders." Abraham Farissol, *Iggeret Orhot Olam,* Prague, 1793. Annenberg Research Institute, Philadelphia (Cat. no. 168)

Map of Israel and the surrounding areas, accompanying the 1604 edition of Mordecai Jaffe's *Levush HaOrah.* National Library of Canada, Jacob M. Lowy Collection (Cat. no. 165)

Samaritan or Palmyrene scripts, though paleo-Hebrew script, which is closely related to them, is found between the Euphrates and Israel, called Chaldaica. The map of Africa continues this pattern, giving extensive coverage to Ethiopic, also a Semitic language. The Christian provenance of the map is clear from what was chosen as the Hebrew label, a literal translation of the first few words of the Lord's prayer.

Abraham ben Mordecai Farissol [Peritsol] of Ferrara and Mantua (1452–ca. 1528)

אגרת ארחות עולם וכו'

[Iggeret Orhot Olam, Iggeret Teiman Petah Tiqvah, Yesod Moreh VeSod Torah]
Prague: 1793
Annenberg Research Institute, Philadelphia

168

*Farissol came from a distinguished Avignon family, and his cultural environment was steeped in Jewish learning and the sciences. He included comments on botany, zoology, cartography, natural history, astronomy, astrology, medicine, and similar subjects in his writings, but was not an innovative contributor to them. He served for much of his life as a copyist of Hebrew books, but also assumed important (though not always appreciated) roles as a teacher, cantor, and community spokesman. He wrote commentaries on the mishnaic tractate *Pirqei Avot* and the biblical books of Job and Ecclesiastes.

Participation in Ferrara's interfaith debates in the late 1580s led Farrisol to compose *Magen Avraham*, a source book and guide for such situations. Its popularity is evidenced by its existence in at least thirty-four manuscripts and by the fact that, like many other seminal works – for example, Rashi's Torah and Talmud commentaries, Rav Sherira Gaon's *Iggeret*, and *Seder Rav Amram Gaon* – it has been augmented freely.

Farissol's most famous work, *Iggeret Orhot Olam*, contains much general geographic information: descriptions of the seven climatic zones, the continents, and the nature of latitudinal and longitudinal divisions of the earth; a chapter on the ten lost tribes that includes much information on David Reuveni; and a mixture of geographically based folklore, highlighted by an attempt to locate the Garden of Eden.

According to D. Rudermann, author of the most detailed study on Farissol and his work, almost half of *Iggeret Orhot Olam* was borrowed from Francanda Montalboddo's *Paesi novamente retrovati e novo mundo da Alberico Vesputio fiorentino intitulato*, essentially a compendium of all known information about New World discoveries published in 1507 and in at least a dozen editions in three different languages. Other works were also collated in the process of composing *Iggeret Orhot Olam*, though it is far from just a collection or translation of earlier writings.

Iggeret Orhot Olam was translated into Latin and annotated by Thomas Hyde; it has appeared in at least nine Hebrew editions, of which Hyde's was the most influential. It was re-edited and supplemented with discussions and illustrations of, among other things, skindiving in Calcutta and whaling (Prague, 1793) by Israel Landau (son of Rabbi Ezekiel Landau), an important *Haskalah* figure in that city. The Hebrew title page identifies him only as "I.L. of the House of Levi."

Benjamin [ben Jonah] of Tudela (late twelfth century)

מסעות של רבי בנימין

[Masaʿot Shel Rabbi Binyamin]
Ferrara: Abraham ibn Usque, 1556
National Library of Canada, Ottawa (Jacob M. Lowy Collection)

169

Jewish travel is as ancient as Judaism itself. Abraham, biblical father of the Jewish people, began life in Mesopotamia, moved westward to what was later called Israel and southward to Egypt, and finally settled in Israel. Many other biblical characters were equally mobile, and later Jews have followed this pattern. Their voluntary or forced migrations have led them to live in, contribute to, populate, and develop virtually every area inhabited by humans. Ancient letters and documents, as well as the travels of individuals provided insight into the lifestyles of Jews who lived in exotic places, and occasional references tell of journeys to visit them.

Benjamin of Tudela was far from the first medieval Jewish traveler, but the record of his journey has become the most popular Hebrew travelogue ever written. Benjamin's travels took him from Spain to France, Italy, Greece, Egypt, Turkey, Armenia, Syria, Israel, Iraq, Iran, Arabia, Yemen, Tibet, Ethiopia, India, Germany, and Russia. He visited some areas casually, and explored others thoroughly. His reports are full of interesting information on the sizes of the various Jewish communities he encountered and their leaders. His comments on local institutions, demographics, and customs are among the only medieval reports of some isolated communities that have reached us.

The different angles of vision from the deck and above it. Tobias Cohn, *Ma'aseh Tuviah*, Venice, 1708. National Library of Canada, Jacob M. Lowy Collection (Cat. no. 73)

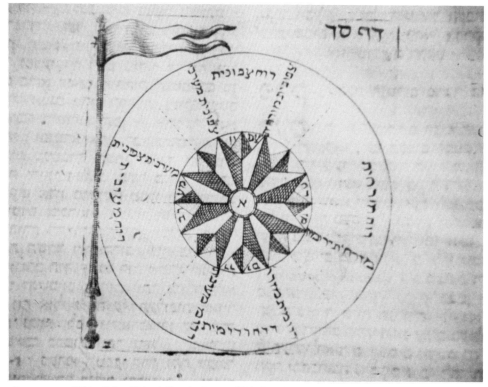

Compass rose. Tobias Cohn, *Ma'aseh Tuviah*, Jesnitz, 1721. Jewish Public Library, Montreal (Cat. no. 74)

Psalm Scroll from Masada. The Israel Museum, The Shrine of the Book, Jerusalem (Cat. no. 4)

Sections 20 and 21 of Al-Farghani's *Kitzur Almagest*. Hebrew Union College, Cincinnati, MS #891.3 (Cat. no. 11)

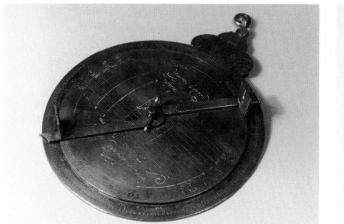

Planispheric Astrolabe: Used to reproduce stellar configurations of the sky at any moment in time. From it may be deduced the time, the rising and setting of a specific star, and the length of day and night.
Brass. France, 16th century.
The David M. Stewart Museum Collection. Montreal

Introduction to Isaac Israeli's *Yesod Olam*. Biblioteca Palatina, Parma, MS #3165 (Cat. no. 23)

Nocturlabe: An astronomical instrument used to measure time at night, using the moon and the stars. Brass and silver. France (?), Sixteenth century. The David M. Stewart Museum Collection. Montreal.

Telescope: Brass, leather, ivory. Made by Passement, ingénieur du Roy. Paris, France. Eighteenth century. The David M. Stewart Museum Collection, Montreal.

Top
Isaac ben Solomon ben Alhadib, *Sefer Orah Selulah*, bound with Abraham Bar Hiyya's *Hokhmat HaHizzayon*, Part I, *Heshbon Mehalekhot HaKokhavim*. The Vatican, MS Ebr. #379 (Cat. no. 14)

Jacob's Staff (Cross-staff): A land-surveying instrument that became very useful at sea during the Middle Ages. It was used to measure the height of the sun above the horizon. It was used by sixteenth-century explorers.
Wood, brass, ivory. 1726.
The David M. Stewart Museum Collection. Montreal.

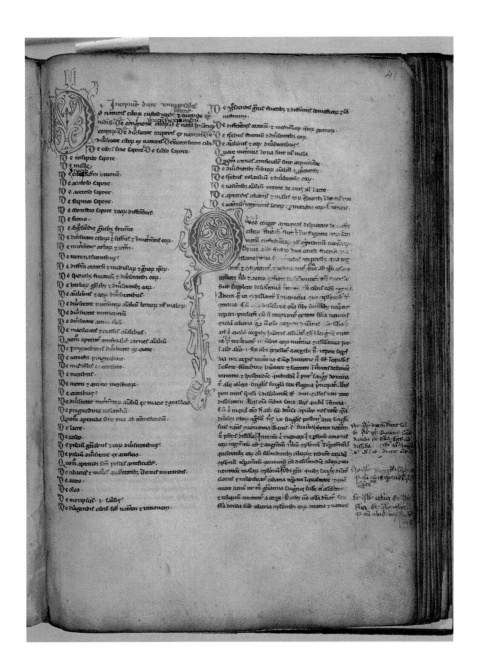

Top
The first passage of Isaac Judaeus' *Diete universales* and *Diete particulares*. McGill University, Montreal, Osler Library of the History of Medicine #7626 (Cat. no. 85)

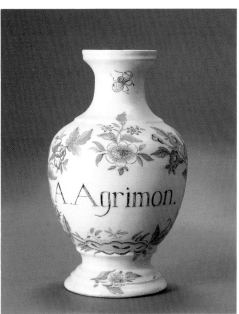

Medicine Vase: Polychromed glazzed earthenware. Bordeau, France, Circa 1750.
The David M. Stewart Museum Collection. Montreal.

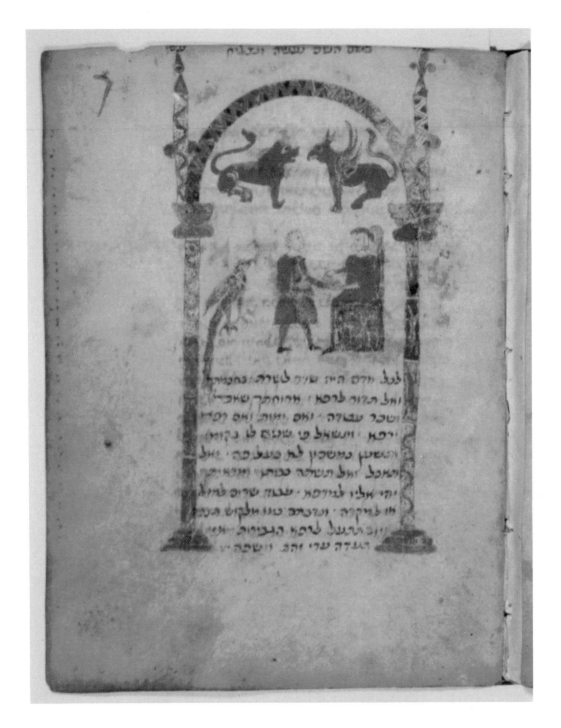

Top
Bloodletting, Bruno de Lungoburgo,
Sefer HaKeritot.
The Vatican, Ebr. #462
(Cat. no. 98)

Bloodletting Bowl: Used to collect blood. Bloodletting was a commonly used procedure to cure all kinds of sicknesses; often, the result was greater infection.
Blue camaieu glazed earthenware.
Nevers, France, Circa 1730.
The David M. Stewart Museum Collection, Montreal.

A fifteenth century physician symbolizing Avicenna. Biblioteca Universitaria, Bologna, MS #2297 (Cat. no. 88)

Perpetual Calendar: Pocket instrument used to determine not only the day and the month, but also the length of the day and the night.
Silver, gilt bronze, semi-precious stones. Great Britain. Eighteenth century.
The David M. Stewart Museum Collection. Montreal.

Microscope: Wood, cardboard, glass.
Germany, late 17th century.
The David M. Stewart Museum Collection.
Montréal.

Water Fountain: This laboratory apparatus was used to demonstrate the effects of the atmospheric pressure on liquids.
Brass, lacquered "à la chinoise." Father Nollet physics cabinet, Dijon, France. circa 1745.
The David M. Stewart Museum Collection, Montreal.

The balance, held here by an armed man, frequently symbolized careful preparation of the calendar. Hebrew Union College, Cincinnati, MS #903. (Cat. no. 143)

Water Clock: The first waterclocks appeared in ancient Egypt and worked on the same principle as the hourglass.
Wood, brass, glass, cork, iron. York, Great Britain, 1614.
The David M. Stewart Museum Collection, Montreal.

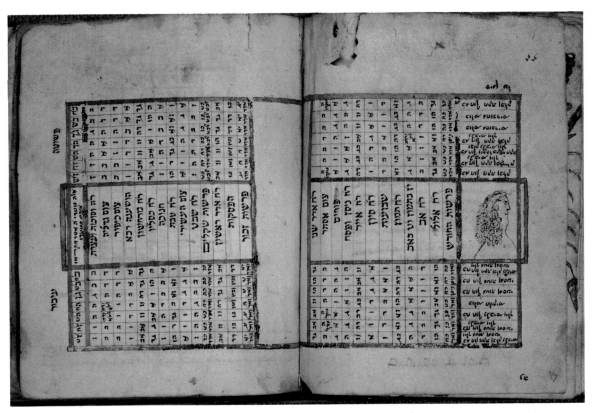

A chart indicating the days of the week on which Rosh Hodesh (the new moon) and Jewish holidays fall in regular and leap years. Hebrew Union College, Cincinnati MS #906. (Cat. no. 142)

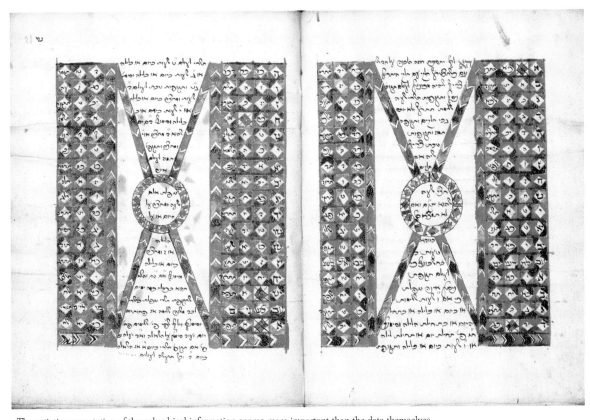

The artistic presentation of the calendrical information seems more important than the data themselves. Hebrew Union College, Cincinnati, MS #903. (Cat. no. 143)

A nature scene accompanies a Hebrew mnemonic for remembering the days of the week on which various pairs of holidays regularly fall. Hebrew Union College, Cincinnati, MS #906. (Cat. no. 142)

The upper panel depicts Absolom caught in a tree by his hair and Joab attacking him with a lance (*II Samuel* 18:9ff); the lower one contains a selection of animals, including a unicorn. Hebrew Union College, Cincinnati, MS #906. (Cat. no. 142)

A wheel of the zodiac on which are placed the seven planets, the names of the twelve constellations, and their symbols. Below the wheel are four armed men labelled Pollack, Saracen, Israel, and France. Two of the four seem to be left-handed. Hebrew Union College, Cincinnati, MS #906. (Cat. no. 142)

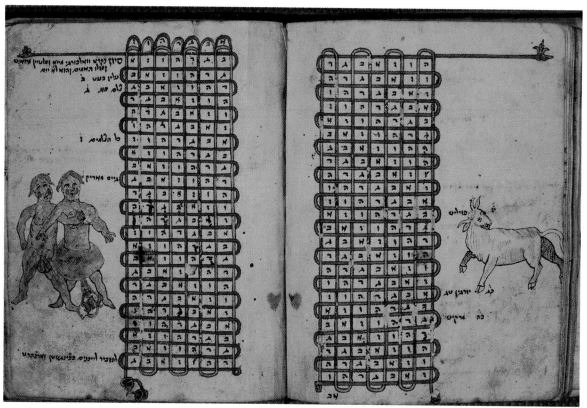

A calendrical chart for the month of Sivan (May–June). To the right is Taurus to the left Gemini. Hebrew Union College, Cincinnati, MS #906. (Cat. no. 142)

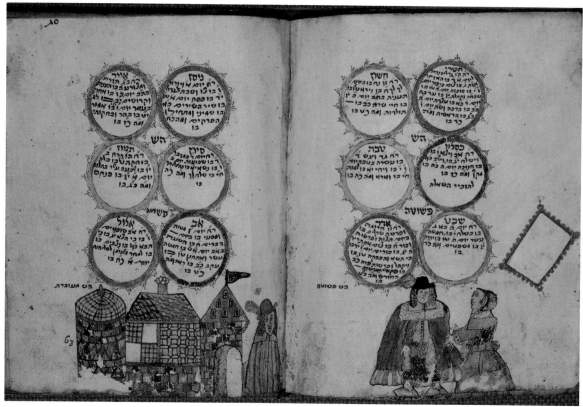

A local house and its inhabitants accompany twelve circles, each containing the name of one month in the year and the dates of Rosh Hodesh and the holidays, as well as the weekly Torah readings for the month. Hebrew Union College, Cincinnati, MS #906. (Cat. no. 142)

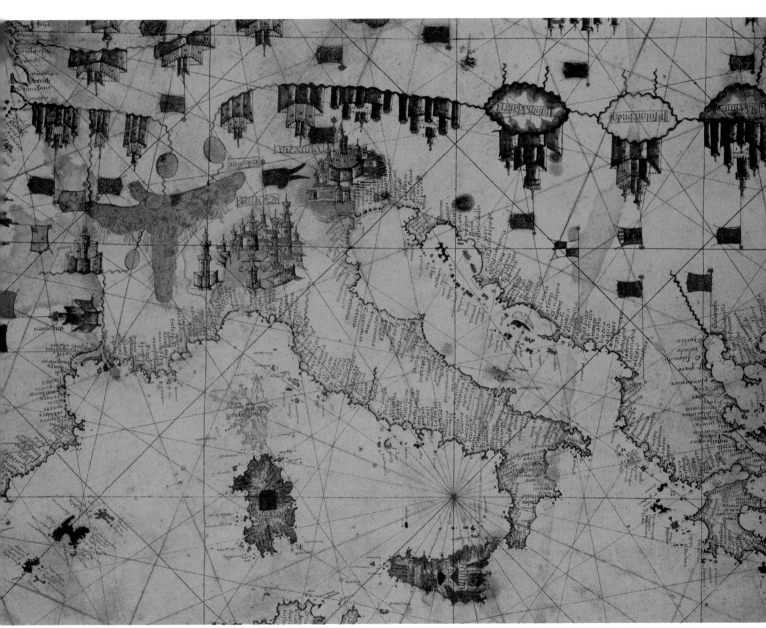

Italy from Judah ibn Zara's, "Map of the Mediterranean," 1497. The Vatican, Borgiano VII (Cat. no. 161)

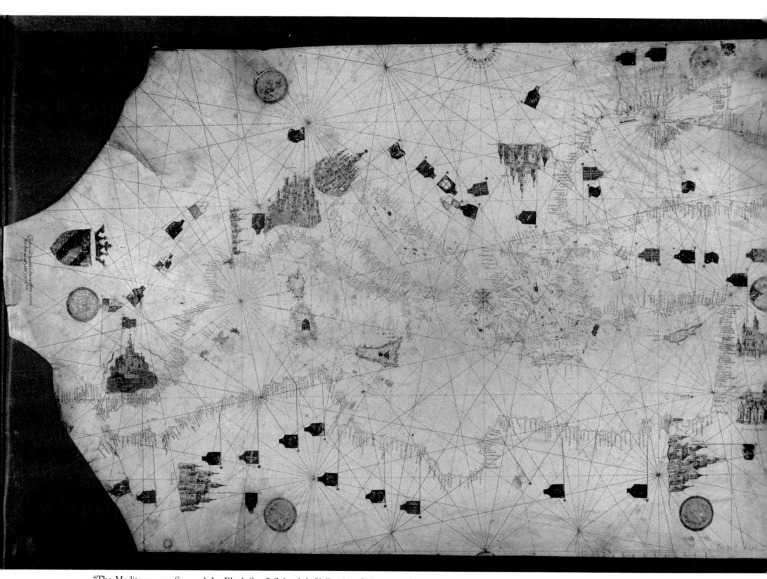

"The Mediterranean Sea and the Black Sea," Gabriel de Vallsecha of Majorca, 1447. Bibliothèque Nationale, Paris, Res. Ge C #4607 (Cat. no. 160)

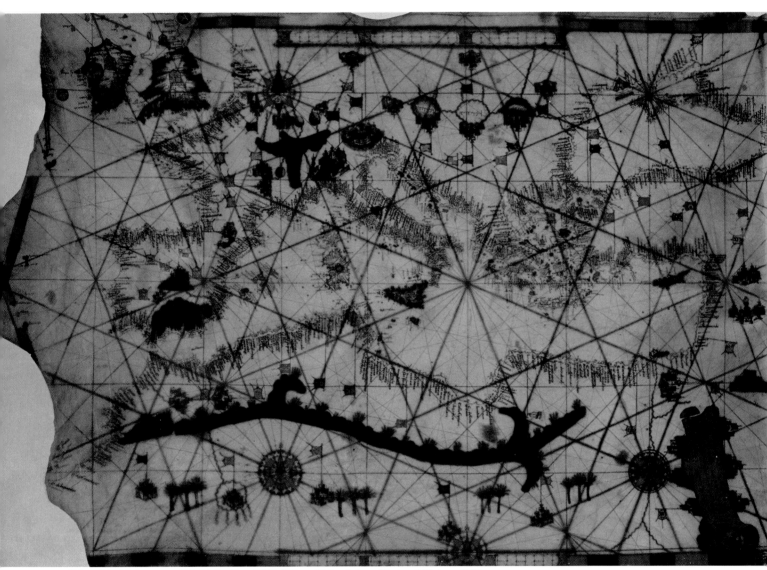

"Map of the Mediterranean," Judah ibn Zara, 1500. Hebrew Union College, Cincinnati (Cat. no. 161)

The angle of vision from the crow's nest of a ship to a castle across the curvature of the earth. Moses ben Barukh Almosnino of Salonika, *Peirush Kadur HaOlam* and *Shaʿar HaShamayim,* Biblioteca Palatina, Parma, MS #3037 (Cat. no. 33)

Pocket Globe: C. Price. London, Great Britain. Circa **1755**. David M. Stewart Museum, Montreal

Postscript

By the nineteenth century, the human race had made great strides toward understanding Job's whirlwind and a myriad of other natural phenomena, but was no less awestruck by their presence nor less helpless to respond to many of them. Confrontation with natural forces that we are incapable of thwarting or controlling underscores our frailty and our inability, despite constant technological advances, to progress far beyond the attitude of Job.

In many ways this merely re-emphasizes human weakness, even – perhaps especially – with respect to forces that we ourselves have set in motion. And many frustrated people have responded by retreating from science and discovering anew the appeals of undisturbed nature and/or naive religion. Lately, many Jews have moved in the direction of pan-entheism, a theology that derives from an interpretation of Isaiah 6:3, "The whole world is full of His glory," taken to suggest the presence of divinity in all that exists. Though it may offer an inspiring response to contemporary materialism, it also bears similarities to the pantheism of earlier eras.

God challenged Job to understand nature and to participate actively and creatively in it. The successes of modern scientists in furthering our understanding of the world – including exploring the sub-atomic particles of which everything is composed and deciphering the codes that control human genetics – are truly impressive and demonstrate that we have begun to meet the challenge. Progress in the battle against cancer and the successful launch of an American probe destined for Neptune are typical additions to a constantly growing list of recent scientific accomplishments, some of which enable us to participate actively in nature by creating new life forms and building machines to do what our physical limitations prevent us from doing.

But we remain no less in awe than were our ancestors of the natural forces that affect our lives. The secular world has, to some extent, depersonalized them, for the ancients might have described them as "gods attacking us," rather than as "forces affecting our lives," but the situations are hardly different. When Hurricane Hugo recently destroyed several Caribbean islands and much of the southeastern United States coastline, modern augurs, called meteorologists, predicted the disaster with great accuracy but were helpless to do more than call for evacuation and *post facto* support. Modern counterparts of Job's friends may be more accomplished at sharing the tragedy and supporting the victim, but they are still unable to prevent his problems.

The hourly reports of aftershocks and devastation resulting from the recent earthquake that shook San Francisco jarred us for days. Modern soothsayers are less able to predict precisely the arrivals of future earthquakes than of hurricanes, though they speak often of the ultimate destruction of the California area through natural causes. Despite the potential analogy, few, other than the most zealous preachers, would liken the situation to the destruction of Sodom, which, according to the Bible, resulted from the evil deeds of its inhabitants. Nevertheless, like Abraham, we may save people from the disaster, but we are powerless to prevent it.

Long ago, Elijah eschewed associating God with storms and earthquakes, but he did not imply their total independence from divine control, either. We remain unable to control these forces, but all save the most devoted atheists look for some glimmer of significance that can justify, however weakly, their impact. The modern explanation of *why* an earthquake or a volcano destroys a town is identical to a description of *how* it does so. But, despite the suggestion's obvious naiveté, who is to say that personal improvement motivated by the belief that human ills result from divine dissatisfaction is not more beneficial to society than the detached fatalism of our modern world?

Despite the magnitude of a myriad of physical forces that confront and thwart us, many scientific successes can be described only in superlatives. Not only has the ability to treat what were once fatal diseases lengthened and improved the quality of life, it has virtually given life to the dead, a power once limited to the deity. The ability to diagnose and cure previously baffling sicknesses, to attach severed limbs to the body and render them functional, to replace worn-out human organs, and, in some cases, even to create replacement parts were only god-like fantasies to our grandparents. Many of our ancestors did believe in *qefitzat haderekh*, the ability to travel vast distances in very short times, but none experienced it. And could our forebearers have even dreamed of the long-distance communication systems we use, which let us project sound and pictures anywhere in the world and send written messages through the air. Could they have fathomed the invention of automobiles or photographic equipment or computers?

The ancients coped with infertility by having the wife provide her husband with a concubine. Now, we can treat it by fertilizing a woman's ovum in a test tube and implanting it, or the naturally fertilized ovum of another woman, in her body; cloning may soon become a serious alternative. They tried to make rain through sympathetic magic; we can seed clouds. They used signal fires; we have satellite-based communication systems.

The human race has contributed much to improving the world through science – and Jews have assumed leadership roles in many of these efforts (as witnessed by over four dozen Jewish Nobel Prize winners in Medicine and Physiology, Chemistry, and Physics, and by many thousands of other contributions). Frequently, however, we have failed to improve our lot without simultaneously destabilizing the natural order in which we thrive. Chemical pollution of the land, air, and water, medicinal poisoning that is an accepted part of prescribed treatments, the devastation of uncontrolled radiation created for positive reasons, and epidemic abuse of synthetic chemicals are all by-products of our scientific achievements. Moreover, the legal and moral implications of our scientific accomplishments – positive or negative – often baffle the mind and call into question the entire creative enterprise. We have learned much, but now, in addition to everything else, we must learn that playing God is accompanied by responsibilities of divine proportions.

Study the world and what is in it, God told Job, and then try to share in My creation. When you can do that, you will be ready to talk about understanding and perhaps improving human fate. "The heavens are the Lord's, but he has given the earth to mankind," declared an inspired psalmist (Psalm 115:16). The heavens surely declare the glory of the Lord, but what does the state of the earth say about us?

A picture of a whirlwind. Tobias Cohn, *Ma'aseh Tuviah*, Venice, 1708. National Library of Canada, Jacob M. Lowy Collection (Cat. no. 73)

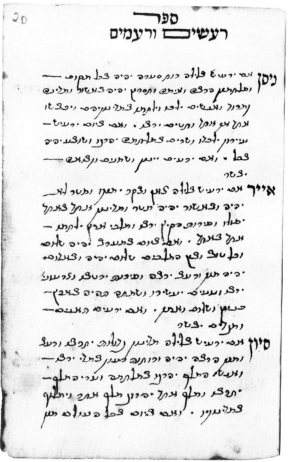

Sefer Ra'ashim VeRa'amim, The Book of Earthquakes and Thunders. The Wellcome Institute Library, London MS Heb. A 17 (Cat. no. 80)

Glossary

Anti-Sabbatian. Those opposed to the teachings of Sabbetai Tzvi, a seventeenth-century messianic figure.

Ashkenaz, Ashkenazic. The customs and rites of the Jews of central and eastern Europe, particularly as contrasted with those of the Sefardic communities.

Derush. Rabbinic homiletics, particularly applied to the Torah.

Gaon. Rabbinic leader of the Babylonian Jewish community from approximately 800–1200 CE. The term was sometimes applied to later rabbis as well.

Halakhah, halakhic. The system of Jewish religious law.

Halitzah. Ceremony in which a childless woman refuses to marry her deceased husband's brother.

Haskalah. The Jewish enlightenment of the past several centuries.

Kabbalah. Jewish mysticism, particularly as taught during the Middle Ages and in the Zohar.

Polyglot. Multi-lingual, particularly as applied to editions of the Bible that contain the text in many languages.

Sefirot. The ten symbolic emanations of the divine presence that underlie much Kabbalistic theology and lore.

Sefarad, Sefaradic. The customs and rites of the Jews of the Iberian Peninsula, North Africa, Yemen, and elsewhere, particularly as contrasted with those of the Ashkenazic communities.

Talmud, Palestinian, and Babylonian. The rabbinic discussions about the Mishnah and related biblical and religious concerns that are the basis of Jewish religious thought and practice; completed approximately 1,500 years ago.

Torah. The Pentateuch, the first five books of the Bible.

Tosefta. "Supplement"; a companion text to the Mishnah that contains early rabbinic legal material and Bible interpretations.

Tzitzit. Tassels prescribed by the Bible to be worn by males on the corners of their garments and now part of the *tallit* or prayer shawl worn during daily worship.

Zohar. The most seminal medieval work of Jewish mysticism.

Bibliography

GENERAL REFERENCE
Gibb, H.A.R., et. al., eds. *The Encyclopaedia of Islam*. New Edition. Leiden: E.J. Brill, 1986.
Roth, C. and Wigoder, G., eds. *Encyclopaedia Judaica*. Jerusalem: Keter, 1972.
Singer, Isidore. *The Jewish Encyclopedia*. New York: Funk and Wagnalls, 1901.
Waxman, Meyer. *A History of Jewish Literature*. New York: Thomas Yoseloff, 1960.
Zinberg, Israel. *A History of Jewish Literature*. Vols. 1-3. Philadelphia: Jewish Publication Society, 1972-3. Vols. 4-12. New York: Ktav, 1974-78.

CATALOGUES
Allan, Nigel. "Catalogue of Hebrew Manuscripts in the Wellcome Institute, London." *Journal of Semitic Studies 27 (1982)*.
Allony, N. and Levinger, D.S. *Reshimat Kitvei-HaYad HaIvriim BaMakhon*. Vol. 3, *Kitvei HaYad SheBeSifriyat HaVatican*. Jerusalem: Rubin Mass, 1968.
Francis, W.W., Hill, R.H., and Malloch, A., eds. *Bibliotheca Osleriana*. Montreal: McGill-Queens University Press, 1969.
Antonucci, Gaetano Zaccaria. *Serie Di Opere Ebraiche Impresse dei Celebri Tipografi Soncini* ... 1870. Reprint. Bologna.
Assemanus, Stephanus. *Bibliothecae Apostolicae Vaticane Codicum Manuscriptorum Catalogus ... Rome: 1755.*
Busi, Guilio. *Edizioni Ebraiche Del XVI Secolo Nelle Biblioteche Dell'Emilia Romagna*. Bologna: Edizioni Analisi, 1987.
De Rossi, I.B. *Mss. codices hebraici biblioth. I. B. de-Rossi* ... Vols. 1-3. Parma: 1803.
Friedberg, C.B. *Bet Eked Sepharim* ... Second Edition. Tel Aviv: 1951.
Hill, Brad Sabin. *Incunabula, Hebraica & Judaica: Five Centuries of Hebraica and Judaica, Rare Bibles, and Hebrew Incunables from the Jacob M. Lowy Collection*. Ottawa: National Library of Canada, 1981.
Hirschfeld, Hartwig. *Descriptive Catalogue of the Hebrew MSS. of the Montefiore Library*. London: MacMillan, 1904.
Ishkandar, A.Z. *A Catalogue of Arabic Manuscripts on Medicine and Science in the Wellcome Historical Medical Library*. London: The Wellcome Historical Medical Library, 1967.
Maher, Paul. *The History of Science: A Collection of Manuscripts from the Library of the Jewish Theological Seminary — An Index to the Microfilm Collection, Reels 1-17*. Ann Arbor, Michigan: University Microfilms International, 1980.
Marx, Moses. "A Catalogue of the Non-Hebrew Books Printed in the Fifteenth Century Now in the Library of the Hebrew Union College." *Studies in Bibliography and Booklore* 5 (1961).
Mendelsohn, Isaac. *Descriptive Catalogue of Semitic Mss. (Mostly Hebrew) in the Libraries of Columbia University*. Unpublished.
Modona, Leonello. *Catalogo Dei Codici Ebraici della Biblioteca della R. Universita Di Bologna*. Firenze: 1889.
Munk, S, Derenbourg, J., and Zotenberg, H. *Catalogues des manuscrits hebreux et Samaritains de la Biblioteque Imperiale*. Paris: 1866.
Neubauer, Adolf and Cowley, A.E. *Catalogue of the Hebrew Manuscripts in the Bodleian Library* ... Vols. 1-2. Oxford: Clarendon Press, 1886-1906.
Steinschneider, Moritz. *Die Hebraeischen Uebersetzungen des Mittelalters und die Juden als Dolmetscher*. 1893. Reprint, Graz: Akademische Druck- u. Verlagsanstalt, 1956.
Strauss, B. *Ohel Barukh - The B. Strauss Library: The Books in Hebrew Characters*. London: Shapiro Vallentine, 1959.
Tamani, Giuliano. *Inventario dei manoscritti ebraici di argomento medico della biblioteca Palatina di Parma*. Firenze: 1967.

VARIA
Abramson, Shraga. *Tractate Àbodah Zarah of the Babylonian Talmud: Ms. #44830 Jewish Theological Seminary of America*. New York: The Jewish Theological Seminary of America, 1957.
Alexander Altmann. *Moses Mendelssohn: A Biographical Study*. Philadelphia: Jewish Publication Society, 1973.
Al-Farghani, Ahmad ben Muhammad [ben Kathir]. *Jawamiʿ ʿim al-nujum wa-usul al-harakat al-samawiya*. Frankfort am Main: J.W. Goethe Universitaet, 1986.
Al-Hassan, Ahmad Y. and Hill, Donald R. *Islamic Technology*. Cambridge: Cambridge University Press, 1986.
Allan, Nigel. "Illustrations from the Wellcome Institute Library: A Polish Rabbi's Circumcision Manual." *Medical History* 33 (1989): 247-254.
Alter, Jiri (George). "Two Renaissance Astronomers: David Gans and Joseph Delmedigo." *Rozpravy Ceskoslovenske Akademie* 68, no. II (Prague: 1958): 45-75;
Amram, David Werner. *The Makers of Hebrew Books in Italy: Being Chapters in the History of the Hebrew Printing Press*. London: Holland Press, 1963.
Arano, Luisa Cogliati. *The Medieval Health Handbook Tacuinum Sanitatis*. New York: George Braziller, 1976.
Baron, Salo Wittmayer. *A Social and Religious History of the Jews*. Vol. 8, *History and Science*. New York: Columbia University Press, 1958.
Barzilay, Isaac. *Yoseph Shlomo Delmedigo (Yashar of Candia): His Life, Works and Times*. Leiden: E.J. Brill, 1974.
Benayahu, Meir. "*Beit Defuso Shel Yira'el Bak BeTzefat VeReishit HaDefus BiYerushalayim.*" *Areshet* 4 (1966): 271-95.
Ben-Sasson, Jonah. *Mishnato HaIyyunit Shel HaRama*. Jerusalem: Israel Academy of Sciences and Humanities, 1984.
Ben-Zvi, Izhak. *Sefer HaShomeronim*. Revised edition. Jerusalem: Yad Yitzhaq Ben Zvi, 1970.
Bologna, Giulia. *Illuminated Manuscripts: The Book Before Gutenberg*. London: Thames and Hudson, 1988.
Castiglioni, Artura. "The Contribution of the Jews to Medicine." In *The Jews: Their Role in Civilizaton*, edited by Louis Finkelstein. Fourth edition. New York: Schocken, 1971.
Coyne, G.V., Heller, M., and Zycinski J., eds. *The Galileo Affair: A Meeting of Faith and Science*. Citta del Vaticano: Specola Vaticana, 1985.
— eds. *Newton and the New Direction in Science*. Citta del Vaticano: Specola Vaticana, 1988.
De Hamel, Christopher. *A History of Illuminated Manuscripts*. Boston: David R. Godine, 1986.
Dodge, Bayard. *The Fihrist of al-Nadim*. New York: Columbia University Press, 1970.
Doresse, Jean. *The Secret Books of the Egyptian Gnostics*. New York: Viking Press, 1960.
Espenshade. H. "A Text on Trigonometry by Levi ben Gerson." *Mathematics Teacher* 60 (1967): 628-37.
Feldman, W.M. *Rabbinical Mathematics and Astronomy*. New York: Herman Press, 1965.
Fischel, Walter J. "Garcia de Orta - A Militant Marrano in Portuguese-India in the 16th Century." *Salo Wittmayer Baron Jubilee Volume*, English Section, Volume I. Jerusalem: American Academy for Jewish Research, 1974.
Fishman, David Elliot. *Science, Enlightenment and Rabbinic Culture in Belorussian Jewry 1772-1804*. Ph.D. Diss., Harvard University, 1985; University Microfilms International.
Forbes, R.J. *Studies in Ancient Technology*. Vols. 1-9. Second edition. Leiden: E.J. Brill, 1964.
Friedlaender, Jonathan und Jakob Kohn. *Maase Efod*. Vienna, 1865.
Friedenwald, Harry. *The Jews and Medicine: Essays*. 1944. Reprint. New York: Ktav, 1967.
Gandz, Solomon. *Studies in Hebrew Astronomy and Mathematics*. New York: Ktav, 1970.
Gershenfeld, Louis. *The Jew in Science*. Philadelphia: Jewish Publication Society, 1934.
Glenisson Jean, ed. *Le Livre au Moyen Age*. [France]: Presses du CNRS, 1988.
Goldstein, Bernard R. "Astronomical and Astrological Themes in the Philosophical Works of Levi ben Gerson." *Archives Internationals d'Histoire des Sciences*, (1976): 221-4.
— *The Astronomical Tables of Levi Ben Gershon*. New Haven, Connecticut: The Connectivut Adademy of Arts and Sciences, 1974.
— *The Astronomy of Levi ben Gerson (1288-1344): A Critical Edition of Chapters 1-20 with Translation and Commentary*. New York: Springer, 1980.
— "Levi ben Gerson's Analysis of Precession." *Journal for the History of Astronomy* 6 (1975): 31-41.
— "The Medieval Hebrew Tradition in Astronomy." *Journal of the Americal Oriental Society* 85 (1965): 145-148.
— "Preliminary Remarks on Levi ben Gerson's Contributions to Astronomy." *Proceedings of the Israel Academy of Scienes and Humanites* 3 (1969): 239-54.
— "Levi ben Gerson's Preliminary Lunar Model." *Centaurus 18 (1974): 275-88.*
— "Scientific Traditions in Late Medieval Jewish Communities." *Les Juifs au Regard de l' histoire*. (1985): 235-47.
— "The Jewish Contribution to Astronomy in the Middle Ages." postscript to Charles Singer, "Science and Judaism." In *The Jews: Their Role in Civilization*, edited by Louis Finkelstein. Fourth edition. New York: Schocken, 1971.
— "The Hebrew Astrolabe in the Adler Planetarium." *Journal of Near Eastern Studies* 35 (1976): 251-60.
— and Pingree, David. "Horoscopes from the Cairo Geniza." *Journal of Near Eastern Studies* 36 (1977).
Gunther, Robert T. *The Astrolabes of the World* ... Vols. 1-2. 1932. Reprint. London: Holland Press, 1976.
Guttmann, Julius. *Philosophies of Judaism*. Trans. David. W. Silverman. New York: Holt Rinehult and Winston, 1964.

Haberman, A.M. "*Divrei Mavo'.*" *Mishnah im Peirush HaRambam*. Naples, 1492. Reprint. Jerusalem: Mekorot, 1970
Halbronn, Jacques. *Le Monde Juif et l'Astrologie.* Milano: Arche, 1985.
Hantzsch, Viktor. *Sebastian Muenster: Leben, Werk, Wissenschaftliche Bedeutung.* Nieuwkoop: B. De Graaf, 1965.
Harvey, Steven. "Did Gersonides Believe in the Absolute Generation of Prime Matter?" *Jerusalem Studies in Jewish Thought*, 7 (1988).
Henry, John. Review of Andre Neher. *Jewish Thought and the Scientific Revolution of the Sixteenth Century: David Gans (1541-1613), and His Times*. In *The Times Literary Supplement* 478 (5 June 1987): 603.
Idel, M. "Enoch is Metatron." *Jerusalem Studies in Jewish Thought* 6 (1987).
Iskandar, Albert Z. "Ibn Sina, Abu ᶜAli al-Husayn ihn ᶜAbdallah." In *Dictionary of Scientific Biography*. Vol. 15.
Jakobovits, Immanuel. *Jewish Medical Ethics: A Comparative and Historical Study of the Jewish Religious Attitude to Medicine and Its Practice*. New York: Bloch Publishing, 1959.
Jospe, Raphael. *Torah and Sophia: The Life and Thought of Shem Tov Ibn Falaquera*. Cincinnati, Hebrew Union College Press, 1988.
Kahle, Paul. *The Kairo Geniza*. Second edition. Oxford: Basil Blackwell, 1959.
Kennedy, E.S. and Pingree, David. *The Astrological History of Mashaᶜ Allah*. Cambridge: Harvard University Press, 1971.
Kunitzsch, Paul. "On the Authenticity of the Treatise on the Composition and Use of the Astrolabe Ascribed to Messahalla." *Archives Internationales d'Histoires des Sciences* (1981): 42-62.
Kunitzsch, Paul and Fischer, Karl A.F. "The Hebrew Astronomical Codex MS. Sassoon 823." *The Jewish Quarterly Review* 78 (1988).
Lange, Gershon. *Sefer Maassei Choscheb: Die Praxis des Rechners. Ein hebraeisch-arithmetisches Werk des Levi ben Gerschom aus dem Jahre 1321.* Frankfort on Main: Louis Golde, 1909.
Langerman, Tzvi. "*Eimatai Nosad HaLuah HaIvri? Qadmuto Al Pi Hiburo Shel Al-Khwarizmi,*" *Asufot* 1 (1987).
—— "The Making of the Firmament": R. Hayyim Israeli, R. Isaac Israeli and Maimonides." *Jerusalem Studies in Jewish Thought* 7 (1988).
—— *The Jews of Yemen and the Exact Sciences*. (Hebrew). Jerusalem: Misgav Yerushalayim, 1987.
Levey, Martin. "The Encyclopedia of Abraham Savasorda: A Departure in Mathematical Methodology." *Isis* 43 (1952).
—— "Abraham Savasorda and is Algorism: A Study in Early European Logistic." *Osiris* 2 (1954): 50-64.
Levinger, Jacob. "Maimonides' Exegesis of the Book of Job." In *Creative Biblical Exegesis: Christian and Jewish Hermeneutics through the Centuries*, edited by Benjamin Uffenheimer and Henning Graf Reventlow. Sheffield: 1988.
Levy, Raphael. *Astrological Works of Abraham Ibn Ezra*. Baltimore: Johns Hopkins Press, 1927.
Littman, M. *Approaching Infinity: Selected Mathematical Writings of Rabeynu Shlomo of Chelme (Author of Mirkeves Hamishna on the Rambam)*. Monsey, New York: 1989.
Mahler, Raphael. *Hasidism and the Jewish Enlightenment: Their Confrontation in Galicia and Poland in the First Half of the Nineteenth Century*. Trans. by E. Orenstein, A. Klein, and J. Klein. Philadelphia: Jewish Publication Society, 1985.
Malavolti, Mirko. *Medici Marrani in Italia nel XVI e XVII Secolo*. Roma: Istituto Di Storia Della Medicina Dell'Universita' Di Roma, 1968.
Marx, Alexander. "The Scientific Work of Some Outstanding Mediaeval Jewish Scholars." in *Essays and Studies in Memory of Linda R. Miller*, edited by I. Davidson. New York: Jewish Theological Seminary, 1938.
—— "Astrology Among the Jews in the Twelfth and Thirteenth Centuries." In *Studies in Jewish History and Booklore*. New York: The Jewish Theological Seminary of America, 1944.
McCullough, W.S. *Jewish and Mandaean Incantation Bowls in the Royal Ontario Museum*. Toronto: University of Toronto Press, 1967.
Metzger, Therese and Metzger, Mendel. *Jewish Life in the Middle Ages*. New York: Alpine, 1982.
Montgomery, James A. *Aramaic Incantation Texts from Nippur*. Philadelphia, 1913.
Neher, Andre. *David Gans (1541-1613): disciple du Maharal de Prague, assistant de Tycho Brahe et de Jean Kepler*. Paris, 1974. Translated by David Maisel as *Jewish Thought and the Scientific Revolution of the Sixteenth Century: David Gans (1541-1613) and His Times*. Oxford: Oxford University Press, 1986.
Neugebauer, O. *The Exact Sciences in Antiquity*. Second edition. New York: Dover Publishing, 1969.
—— *A History of Ancient Mathematical Astronomy*. Vols. 1-3. New York: Springer-Verlag, 1975.
—— *Astronomy and History: Selected Essays*. New York: Springer-Verlag, 1983. Includes "The Astronomy of Maimonides and Its Sources," originally published in *Hebrew Union College Annual* 22 (1949): 322-63.
Pedersen, Olaf. *A Survey of the Almagest*. Denmark: Odense University Press, 1974.
Pines, Shlomo. *The Collected Works of Shlomo Pines*. Vol. 1, *Studies in Abu'l-Barakat al-Baghdadi: Physics and Metaphysics*. Jerusalem: Magnes Press, 1979. Vol. 2, *Studies in Arabic Versions of Greek Texts and in Medieval Science*. Jerusalem: Magnes Press, 1986.
Pingree, David and Madelung, Wilfred. "Political Horoscopes Relating to Late Ninth Century Alids." *Journal of Near Eastern Studies* 36 (1977).
Rabinovitch, Nachum L. *Probability and Statistical Inference in Ancient and Medieval Jewish Literature*. Toronto: University of Toronto Press, 1973.
Rabbinovicz, R. *Maamar al Hadpasat HaTalmud*. Jerusalem: Mossad HaRav Kook, 1965.
Ring, Malvin E. *Dentistry: An Illustrated History*. New York: Harry N. Abrams, 1985.
Robb, David M. *The Art of the Illuminated Manuscript*. New York: A.S. Barnes, 1973.
Rosensweig, Bernard. *Ashkenazic Jewry in Transition*. Waterloo, Ontario: Wilfrid Laurier University Press, 1975.
Rosner, Fred, ed. and trans. *Julius Preuss' Biblical and Talmudic Medicine*. New York: Sanhedrin Books, 1978.
—— and Muntner, Suessman. *The Medical Aphorisms of Moses Maimonides*. Vols. 1-2. New York: Bloch Publishing, 1973.
Roth, Cecil. *A Life of Menasseh ben Israel*. Philadelphia: Jewish Publication Society, 1934 pp. 132-4.
—— *Gleanings: Essays in Jewish History Letters and Art*. New York: Hermon Press, 1967.
—— *The Jews in the Renaissance*. Philadelphia: Jewish Publication Society, 1964.
Roth, Sol. *Science and Religion*. New York: Yeshiva University, 1967.
Ruderman, David B. *Kabbalah, Magic and Science*. Cambridge: Harvard University Press, 1988.
—— *The World of a Renaissance Jew: The Life and Thought of Abraham ben Mordecai Farissol*. Cincinnati: Hebrew Union College, 1981.
Sarfatti, Gad. B. *Munahei HaMatematiqa BaSifrut HaMadaᶜit HaIvrit Shel Yemei HaBeinayyim*. Jerusalem: Magnes Press, 1968.
Sarton, George. *Appreciation of Ancient and Medieval Science During the Renaissance (1450-1600)*. New York: A.S. Barnes, 1961.
Savitz, H. A. *Profiles of Erudite Jewish Physicians and Scholars*. Chicago: Spertus College of Judaica Press, 1973.
Scholem, G. *Kabbalah*. New York: Quadrangle, 1974.
Schrire, T. *Hebrew Magic Amulets: Their Decipherment and Interpretation*. New York: Behrman House, 1966.
Siev, Asher. *Rabbeinu Moshe Isserles (Rama)*. New York: Yeshiva University, 1972.
Singer, Charles. "Science and Judaism." In *The Jews: Their Role in Civilizaton*, edited by Louis Finkelstein. Fourth edition. New York: Schocken, 1971.
Singleton, Charles S., ed. *Art, Science, and History in the Renaissance*. Baltimore: John Hopkins Press, 1967.
Sirat, Colette. *La Lettre Hebraique et Sa Signification*. Published with Leila Avrin, *Micrography as Art*. Jerusalem: The Israel Museum, 1981.
—— and Beit-Arie, Malachi. *Manuscrits Medievaux en Caractères Hébraïques Portant des Indications de Date Jusqu'à 1540*. Vols. 1-3. Jerusalem: Academie Nationale des Sciences et des Lettres d'Israel, 1972-86.
Solon, Peter. "*The Six Wings* of Immanuel Bonfils and Michael Chrysokokkes." *Centaurus* 15 (1970): 1-20.
Spanier, Ehud, ed. *The Royal Purple and the Biblical Blue, Argaman and Tekhelet: The Study of Chief Rabbi Dr. Isaac Herzog on the Dye Industries in Ancient Israel and Recent Scientific Contributions*. Jerusalem: Keter, 1987.
Steinschneider, M. *Jewish Literature from the Eighth Century*. 1857. Reprint. New York: Hermon Press.
—— *Mathematik bei den Juden*. Reprint. Hildesheim: Georg Olms Verlagbuchhandlung, 1964.
Stitskin, Leon D. *Judaism as a Philosophy: The Philosophy of Abraham bar Hiyya (1065-1143)*. New York: Yeshiva University, 1960.
Thorndike, Lynn. *A History of Magic and Experimental Science*. Vols. 1-8. New York: Columbia University Press, 1923-58.
Toomer, G.J. *Ptolemy's Almagest*. London: Duckworth, 1984.
Touati, Charles. *La Pensée Philosophique et Théologique de Gersonide*. Paris: Les Éditions de Menuit, 1973.
Vallicrosa, Jose M. M. *La Obra Sefer Hesbon mahlekot ha-kokabim (Libro del calculo de los movimientos de los astros) de R. Abraham Bar Hiyya Ha-Bargeloni*. Barcelona: Consejo Superior de Investigaciones cientificas, 1959.
—— *El libro de los fundamentos de las Tablas astronomicas de R. Abraham ibn Ezra, edicion critica, con introduccion y notas*. Madrid: Consejo Superior de Investigaciones Cientificas – Instituto Arias Montano, 1947.
—— *Estudios sobre historia de la ciencia espanola*. Barcelona: 1949.
—— *Nuevos estudios sobre historia de la ciencia espanola*. Barcelona: 1960.
Van der Waerden, B.L. *Science Awakening I: Egyptian, Babylonian, and Greek Mathematics*. Trans. A. Dresden. Princeton Junction: Scholar's Book Shelf, 1988.
Wirszubski, Chaim. *Pico della Mirandola's Encounter with Jewish Mysticism*. Cambridge: Harvard University Press, 1989.
Yadin, Yigael. *Masada: Herod's Fortress and the Zealots' Last Stand*. New York: Random House, 1966.
Yerushalmi, Yoseph Hayim. *From Spanish Court to Italian Ghetto*. New York: Columbia University Press, 1971.
Yovel, Yirmiyahu. *Spinoza VeKoferim Aherim*. Tel Aviv: Poalim, 1988.
Zimmels, H.J. *Magicians, Theologians and Doctors*. London: Edward Goldston and Son Ltd., 1952.

Index

Abracatabra, 90
Abravanel, Isaac, 39
Abudraham, 101
Abul Kasim, 81
Akkadian, 79
Alchemy, 35, 59
Al-Farghani, 21, 26, 36, 58, 118
Algebra, 23, 49
Alguadez, Meir, 82-83
Alhadib, 120
Allegory, 4
Almagest, see Ptolemy
Almoli, Solomon, 93
Almosnino, Moses, 33, 34, 132
Alphabet, 43, 46, 47, 53, 114
Amulets, 84-5, 88-91
Angel of Death, 87
Aramaic, 85-9
Arbaah Turim, see Jacob ben Asher
Archaelogy, 62
Archimedes screw, 58
Astrolabe, 23, 25, 31, 24, 118
Astrology, 25, 31, 22, 128
Astronomical Tables, 26
Astronomy, 20, 31, 46, 94, 96-9, 127
Averroes, 30-31, 54-5, 58
Avicenna, 58, 67, 69, 81, 123
Azulai, Abraham, 36
Azulai, Hayyim Joseph David, 19

Bar Hiyya, Abraham, 18, 20-3, 25, 31, 37, 100-1
Bashyatchi, Elijah, 37, 46
Benjamin of Tudela, 2, 115
Bible, 3, 4, 9-14, 16-7, 25, 30-1, 33-5, 37, 39, 54-5, 57-9, 73, 79, 81, 93, 95, 103, 105, 107, 111-2, 114-5, 117, 126
Bloch, Marcus, 60-1, 82
Bloodletting, 122
Bonet, Jacob, 26
Bonfils, Immanuel, 31
Brahe, 32
Buenting, Heinrich, 112

Calendar, 31, 75, 85, 97-99, 101-3, 123-6, 128,
Cardoso, Isaac, 55
Caro, 34-5, 37, 65, 72-3, 75, 78, 101
Cartography, 109
Charms, 89, 84
Circumcision, 67, 103-7
Cohn, Tobias, 22, 56, 59, 61, 63-4, 92-3, 110, 116, 135
Coin, 62, 63
Columbus, 33
Compass, 116
Comtino, 44-5
Comtino, Mordecai, 25
Copernicus, Nicolaus 33, 59, 61, 100
Costi ibn Luqa, 46
Cures, 82, 83

Da Costa, Emanuel Mendes, 61
Da Gama, Vasco, 33
Da Orta, 77, 82-3
David ben Aryeh, 105
Dead Sea Scrolls, 8, 13, 117
Delmedigo, 20, 28, 58-9, 67
De Pomis, David, 49, 75, 77
De Rossi, Azariah, 63
De Valsecha, Gabriel, 111
Dietary Laws, 104, 106-7
Dreams, vi, 91, 93
Duran, Profiat, 31

Eclipse, 22, 30
Efodi, 31
Eidlitz, Abraham Moses, 48-9
Elements, 63
Elijah ben Moses Gershon, 49
Emanuel ben Jacob, 30
Enlightenment, 5, 37
Enoch, 17
Eruv, 40, 42, 46-7
Eruvin, 41, 42
Euclid, 38, 40-1, 43, 49, 54,
Eyes, 79, 81
Ezekiel, 16

Falaquera, Shem Tov, 71
Fallopias, 6
Farissol, 10-11, 13, 114-15
Finzi, Mordecai, 26
Fish, 48
Folk Medicine, 90
Fountain, 124
Friesenhausen, 49

Galen, 41, 58, 67, 69, 71
Galileo, 33, 59
Gans, David, 33, 34
Gears, 58
Gemini, 128
Gentili, Moses, 107
Geocentric, 22
Geography, 109-10, 129-31
Geometry, 38, 40, 43-4, 46, 49, 70
Gershon ben Solomon, 29, 58
Gersonides, 6, 10-11, 28-9, 31, 37, 41, 55, 58
God, 3, 54, 58, 132
Graphometer, 123
Grapius, Zacharius, 104-5, 107
Gregory *III,* Pope, 74-5, 97

Hasidism, 6
Heida, Moses, 49
Helm, 6
Hemophilia, 6, 67
Henoch, G. 6
Hippocrates, 41, 58, 67, 69
Homann, Johann Baptist, 114
Horowitz, Shabbetai, 46, 47
Human Body, 92-3
Humours, 63

Ibn Ezra, 11, 24-5, 31, 33-4, 36, 43, 45, 48
Ibn Sina, *see* Avicenna
Ibn Waafid, Abdul Mutarrif, 69
Ibronot, 101, 103
Impotence, 78, 89
Irrigation, 58
Isaac ben Solomon Judaeus, 66, 68, 121
Isaac ibn Sid, 26
Isaac Israeli, *see* Isaac ben Solomon
Isaac Israeli, *Yesod Olam,* 36, 119
Isaac Judaeus, *see* Isaac ben Solomon
Isserles, Moses, 27, 34, 36, 100, 106

Jacob ben Asher, 34, 73, 75, 78 101
Jacob ben Mordecai, 91
Jacob Poel, 26
Jacob's Staff, 28, 30, 120
Jaffe, Mordecai, 98, 100, 113-4
Job, 10-12, 15, 30, 73, 132
Jonathan ben Joseph ofRuzhany, 20, 94, 100
Jordan River, 114
Joseph ben Issachar Ber, 114

Josephus, 14, 21
Judah ben Elijah Bunzlau, 43
Judah ben Solomon, 38, 41
Judah Halevi, 52-3
Judah ibn Zara, 112, 129, 131

Karaites, 37, 46, 107
Kepler, 33
Kilaim, 41-2
Kimhi, David, 11
Kuzari, 31, 37, 52-3

Leap Years, 125
Letters, 46-7, 53, 114
Leviticus Rabbah, 65
Levush, see Jaffe
Light, refraction of, 28
Lilith, 88-9
Lindau, Barukh, 61
Lowe, Judah, 34
Lungoburgo, Brunno Da, 73, 122
Lusitanus, Abraham Zacutus, 77
Lusitanus, Rodrigo, 77
Luther, Martin, 71, 70

Maaseh Tuviah, 22, 56, 59, 61, 63-4, 78, 92-3, 110, 116, 135
Magic, 85
Magic and Medicine, 85
Magic Bowls, 85-9
Magnets, 56
Maimonides, 6, 11, 13, 20, 27, 30-1, 37, 40, 42, 45, 54, 57, 62, 67, 69-71, 73, 96-100
Maps, 111, 114, 129-31
Marranos, 5, 55, 61, 83
Masada, 8-9, 14, 117
Mashallah, 21, 31
Mathematics, 37, 39, 48, 59
Measures, 58
Medicines, 55, 61, 80-3, 121
Meldola, David, 103
Mentz, 70
Menz, Abraham, 49
Meteorology, 62, 124, 132-35
Microscope, 124
Midrash Hokhmah, 38
Midwifery, 71, 73
Miracles, 3-4, 37, 54
Mishnah, 40-2, 65, 111
Mishneh Torah, see Maimonides
Mizrahi, Elijah, 25, 34, 46
Montanus, Benedict Arias, 112
Moscato Judah, 31, 53
Moses Ben Maimon, see Maimonides
Moses De Leon, see Zohar
Muenster, Sabastian, 25, 46, 103, 102

Nahmanides, 10, 13, 62
Nansich, Abraham, 78
Nanzig, Abraham, 78
Nature, 3
New Moon, 27, 96-7, 99, 125
Nocturlabe, 119
Noveira, Isaac, 46-7
Numbers, 43, 46-7

Palm Reading, 140
Perpetual Calendar, 123
Philo, 4
Pocket Globe, 132
Poel, 36
Poles, North and South, 110
Polyglot Bibles, 14-6
Portaleone, Abraham, 58
Prescriptions, 76, 82, 84, 90, 121
Psalms, 8, 13, 19, 117
Psalm Scroll, 8, see Masada
Ptolemy, 21, 25-6, 30, 33, 37, 41, 46, 54, 58-9, 111, 20

Qehillot Moshe, 8, 13
Qosta ibn Luqa, 25
Qumran, 13

Rabbinic Bible, 8, 10-3
Raphael ben Jacob Joseph of Hanover, 100
Rashi, 11, 13, 40-1
Regimen Sanitatis, See Maimonides
Ritual Slaughter of Animals, 106-7
Romans, 14, 103

Sacrobosco, 25-6, 34, 46
Sahulah, 22, 27
Sailing, 108, 114, 116, 132
Samaritans, 16-7, 62
Scale, 124
Schick, Barukh, 27, 40-4, 49
Sefer HaMispar, 43
Sefer Ibronot, 101, 103
Sefer Raashim VeRaamim, 134
Sefer Raziel, 88-9
Sefer Yetzira, 53
Shabbetai ben Meir HaKohen, 78
Shadows, 20
Shaduf, 58
Shakh, 78
Shehita, 104, 106-7
Shekel, 62-3
Shema, 89
Shulhan Arukh, 75, 101
Skindiving, 114
Smallpox, 79
Sod Adonai by David ben Aryeh, 105
Solomon, 58
Sphaera Mundi, 100
Spinoza, 17, 50, 55

Tables, 33
Talmud, 8, 19, 37, 40-3, 51, 53, 65, 67, 93, 95
Taurus, 128
Temple, 14, 34, 47, 58
Tobit, 79
Travel, 2, 115
Turim, see Jacob ben Asher
Tuviah ben Meir HaLevi, 43
Tzitzit, 6
Tzurat HaAretz, 23, 25

Unicorn, 126

Vaccination, 79
Vacuums, 56
Vallsecha, 130
Vicinho Joseph, 33
Vilna Gaon, 41
Vital, Hayyim, 35, 53

Wallich, Abraham, 78
Wallich, Judah Leib, 78
Water Clock, 124
Weather, 62
Weights, 58
Weil, Jacob, 104-6
Whirlwind, 14, 135
Wissenschaft, 6, 63
Worms, 60

Yeshiva University, 6
Yesod Olam, 36

Zacuto, Abraham, 27, 32-3, 35-6, 77
Zamosc, Israel, 37
Zodiac, 127
Zohar, 27, 36, 91, 109-11

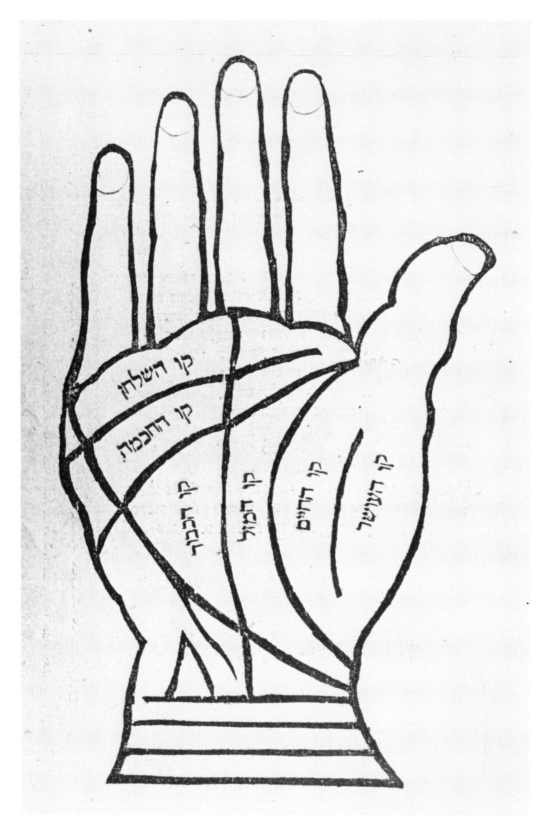

Guide for palm reading. Jacob ben Mordecai of Fulda, *Shoshanat Ya'aqov*. Amsterdam, 1706, Yeshiva University, New York (Cat. no. 131)